Acta Universitatis Upsaliensis

Symposia Universitatis Upsaliensis
Annum Quingentesimum Celebrantis 7

Differential Equations

Proceedings from the Uppsala 1977 International Conference on Differential Equations

Held at Uppsala, Sweden April 18-23, 1977

Uppsala 1977

Edited by G. Berg, M. Essén and Å. Pleijel

Distributor: Almqvist & Wiksell International,
Stockholm Sweden

ISBN 91-554-0698-X

Printed in Sweden by
LiberTryck Stockholm 1977

PREFACE BY THE EDITORS

The Conference was a part of the 500 year anniversary celebrations of the University of Uppsala. It has therefore been natural to include its proceedings in the series published in connection with the Jubilee year which was also advantageous because of the organization established by the University for this purpose. The manuscripts were prepared by the authors themselves which has caused certain inhomogeneities but made a rapid publication possible. In gratitude the volume is dedicated to the University of Uppsala.

The program of the Conference dealt with the theory of differential equations and its connections with different fields as theoretical background. A certain emphasis was put on the deficiency index problem. To cover these aspects a number of plenary speakers were invited, whose lectures are recorded in this volume. The Conference was open to all mathematicians, and in particular to those who wanted to give short communications on new results within the theory of differential equations and related subjects. These communications are given in the proceedings by their titles only.

More than one hundred participants were registered from different parts of the world as can be seen from the list of addresses of the speakers. Many Swedish mathematicians attended the conference as a whole or came to lectures of special interest for them selected from a preliminary schedule sent out beforehand.

The main support to the Conference was given by the University of Uppsala, financially but also in many other respects. The Conference was also supported by the International Mathematical Union and by the Swedish Government through its Educational Department. Aid was given by the Swedish Institute to assist participants from countries with non-convertible currencies which gave the Conference a really international character. The Conference was sponsored by the Royal Swedish Academy of Sciences whereby it could use the exchange affiliations between this

Academy and Academies in other countries. The Organizing Committee expresses its gratitude to all these institutions.

Above all the Committee wish to thank all participants, plenary speakers and others, who made the Conference a success. The Conference would not have been possible without support and help from the Mathematical Department of the University. We thank the staff including colleagues and secretaries.

The Organizing Committee

Gunnar Berg   Matts Essén   Åke Pleijel

CONTENTS

Lectures given at the Conference which are not represented by contributions to these Proceedings.

R.J. Amos:

On an integral inequality of Essén and Keogh concerning a function and its derivative on a finite interval.

J.W. Bebernes:

Invariance and solution set properties for some nonlinear differential equations.

H. Behncke:

Deficiency indices for Schrödinger operators with singular potentials.

C. Bennewitz:

On a theorem by Niessen.

H.E. Benzinger:

A canonical form for ordinary differential operators.

P. Besala:

On the Cauchy problem for second order parabolic equations.

I. Bihari:

On the oscillation of Bôcher's pairs with respect to half-linear differential equations.

C.E. Billigheimer:

Ordinary self-adjoint linear differential and difference operators.

A. Borzymowski:

The uniqueness of solutions of tangential-derivative problems for some partial differential equations of parabolic type.

B.L.J. Braaksma:

Asymptotic expansions for linear difference equations.

P.J. Browne:

Multiparameter problems.

C. Castaing:

Râfle par un convexe aléatoire à variation continue à droite.

W.A. Coppel:

Pseudo-autonomus linear systems.

E.R. Dawson:

Some remarks on the determination of best-possible constants in integral inequalities involving derivatives.

A. Dijksma:

Ordinary differential subspaces in Banach spaces and their adjoints.

A. Elbert:

On solutions of linear second order differential equations.

M. Essén:

On ψ-sequences and the growth of small subharmonic functions.

W.D. Evans:

Essential self-adjointness of powers of Schrödinger and Dirac operators.

M. Faierman:

Multiparameter eigenvalue problems.

J. Fleckinger-Pellé:

Some results in spectral theory for elliptic partial differential operators on unbounded domains.

H. Focke:

Stability of deficiency indices.

C.T. Fulton:

Singular eigenvalue problems with eigenvalue parameter in the boundary conditions.

R.C. Gilbert:

A class of formally selfadjoint ordinary differential operators whose deficiency numbers differ by an arbitrary pre-assigned positive integer.

S. Goldberg:

Characterizations of almost periodic C_0 groups of operators and C_0 semigroups of semi Fredholm operators.

H.S.P. Grässer:

On a variational method for solving boundary value problems of even order.

M.W. Green:

Qualitative properties of the Fitzhugh-Nagumo equation.

P. Habets:

An example of relaxation oscillations.

S.G. Halvorsen:

Bounds for solutions of Sturm-Liouville equations, applied to spectral and integrability conditions.

S. Invernizzi:

Periodic solutions for some differential systems.

E.M. de Jager:

The Schrödinger equation as a singular perturbation.

H. Kalf:

Gauss' theorem and the self-adjointness of Schrödinger operators.

A. Källström:

An integral relation for multiparameter eigenvalue problems.

R.M. Kauffman:

Non self-adjoint eigenfunction expansions and the Weyl theory for some higher order ordinary differential operators.

M. Kisielewicz:

Description of a class of multivalued differential equations with almost weakly stable trivial solutions.

I. Knowles:

The essential self-adjointness of certain differential operators.

K. Kreith:

Systems zeros of fourth order differential equations.

H. Kurss and G. Meyer:

Extensions of a limit-point criterion of Ismagilov.

I. Laine:

Some developments concerning a theorem of J. Malmquist.

A. Lasota:

Differential equations with turbulent solutions.

S.J. Lee:

Perturbation of direct sum differential operators.

N.G. Lloyd:

Periodic solutions and Whitney's theorem.

M. Lopez:

Matrix differential operators.

J.B. McLeod:

The existence of eigenvalues embedded in the continuous spectrum of ordinary differential operators.

H.-D. Niessen:

On the deficiency-index for left-definite eigenvalue problems.

I. Nåsell:

Threshold results in Schistosomiasis.

C.G.C. Pitts:

The divergence of a class of eigenfunction expansions.

S.M. Rankin:

Boundary value problems for Volterra equations related to partial functional differential equations.

T.T. Read:

The asymptotic behavior of solutions of systems of differential equations.

K. Schmitt:

Boundary value problems for infinite systems of differential equations.

I.P. Stavroulakis and Y.G. Sficas:

On the oscillatory and asymptotic behavior of differential equations with deviating arguments.

P. Volkmann:

A perturbation theorem for linear, closed operators acting between Banach spaces.

J. Vosmanský:

Regular monotonicity and differential equations.

J. Walter:

On the definition of the absolute temperature - a reconciliation of the classical method with that of Carathéodory.

K.-O. Widman:

Variational inequalities for vector-valued functions.

A.D. Wood:

The existence of solutions of a non-linear differential equation with preassigned discontinuity.

L.E. Zachrisson:

On an optimality result by M. Essén concerning a coefficient in a simple differential equation.

A. Zettl:

Relations among powers of differential operators in $L^2(0,\infty)$.

Address list of speakers and authors

R.J. Amos:	Mathematics Department, University of Dundee, Dundee, DD1 4HN, Scotland
F.V. Atkinson:	Department of Mathematics, University of Toronto, Toronto, Ontario, Canada
J. Bebernes:	Department of Mathematics, University of Colorado, Boulder, Colorado 80302, USA
H. Behncke:	D-45 Osnabrück, Universität, FB 5, Albrechtstr. West Germany
C. Bennewitz:	Department of Mathematics, University of Uppsala, Sysslomansgatan 8, 752 23 Uppsala, Sweden
E. Benzinger:	Department of Mathematics, 273 Altgeld Hall, University of Illinois, Urbana, Illinois 61801, USA
P. Besala:	ul. Swietojańska 67 m.4, 80-840 Gdansk, Poland
I. Bihari:	Mathematical Institute of the Hungarian Academy of Sciences, 1053 Budapest, Reáltanoda-u.13/15, Hungary
C.E. Billigheimer:	Department of Mathematics, McMaster University, Hamilton, Ontario L8S 4K1, Canada
A. Borzymowski:	Department of Mathematics, Warsaw Technical University, Tanka 37 m.1, 00-355 Warsaw, Poland
B. Braaksma:	Department of Mathematics, University of Groningen, P.O. Box 800, Groningen, The Netherlands
P. Browne:	Mathematics Department, University of Calgary, Calgary, Alberta T2N 1N4, Canada
C. Castaing:	Institut de Mathématiques, Université des Sciences et Techniques du Languedoc, Place Eugène Bataillon, 34060 Montpellier Cedex, France
L. Cesari:	Department of Mathematics, University of Michigan, Ann Arbor, Michigan 48104, USA
E.A. Coddington:	Mathematics Department, University of California, Los Angeles, California 90024, USA

R. Conti: Istituto Matematico "Ulisse Dini", Viale Morgagni 87/A, Firenze I 501 34, Italy

W.A. Coppel: Department of Mathematics, I.A.S., Australian National University, Box 4 P.O., Canberra, Australia

E.R. Dawson: Mathematics Department, University of Dundee, Dundee, DD1 4HN, Scotland

A. Devinatz: Department of Mathematics, Lunt Hall, Northwestern University University, Evanston, Illinois 60201, USA

A. Dijksma: Department of Mathematics, University of Groningen, P.O. Box 800, Groningen, The Netherlands

M.S.P. Eastham: Mathematics Department, Chelsea College, Manresa Road, London SW3 6LX, England

A. Elbert: Mathematical Institute of the Hungarian Academy of Sciences, 1053 Budapest, Reáltanoda-u.13/15, Hungary

M. Essén: Department of Mathematics, Royal Institute of Technology S - 100 44 Stockholm, Sweden

W.D. Evans: Pure Mathematics Department, University College, Cardiff, CF1 1XL, Wales

W.N. Everitt: Mathematics Department, University of Dundee, Dundee, DD1 4HN, Scotland

M. Faierman: Department of Mathematics, Ben Gurion University of the Negev, Beer Sheva, Israel

J. Fleckinger-Pellé: Department of Mathematics - 1R2 -, Université Paul Sabatier, 118 Rte de Narbonne, F-31077 Toulouse Cedex, France

H. Focke: D-45 Osnabrück, Universität, FB 5, Albrechtstr. West Germany

A. Friedman: Department of Mathematics, Lunt Hall, Northwestern University, Evanston, Illinois 60201, USA

C.T. Fulton: Department of Mathematics, Pennsylvania State University, 215 McAllister Building, University Park, Pennsylvania, USA

R.C. Gilbert: Mathematics Department, California State University, Fullerton, California 926 34, USA

S. Goldberg: Department of Mathematics, University of Maryland, College Park, Maryland 20742, USA

H.S.P. Grässer: Department of Mathematics and Applied Mathematics, University of South Africa, P.O.Box 392, Pretoria 0001, South Africa

M.W. Green: Mathematics Department, University of Dundee, Dundee, DD1 4HN, Scotland

P. Habets: Institut Mathématique, Chemin du Cyclotron 2, 1348 Louvain-la-Neuve, Belgium

J. Hale: Division of Applied Mathematics, Brown University, Providence, R.I. 02912, USA

S.G. Halvorsen: Institute of Mathematics, University of Trondheim - NTH, 7034 Trondheim - NTH, Norwey

E. Hille: 8862 La Jolla Scenic Drive N., La Jolla, California 920 37, USA

S. Invernizzi: Istituto di Matematica dell' Università, Piazzale Europa 1, I - 34100 Trieste, Italy

E. de Jager: Mathematical Institute, University of Amsterdam, Roeterstraat 15, Amsterdam, The Netherlands

H. Kalf: Institut für Mathematik, RWTH Aachen, Templergraben 55, D 51 Aachen, West Germany

R.M. Kauffman: Department of Mathematics, Western Washington State College, Bellingham, Washington 98225, USA

Y. Kisielewicz: Institute of Mathematics and Physics, ul. Akademicka 10, 65-240 Zielona Góra, Poland

I.W. Knowles: Department of Mathematics, University of the Witwatersrand, Ian Smuts Avenue, Johannesburg, South Africa

K. Kreith: Mathematics Department, University of California, Davis, California 95616, USA

H. Kurss: Department of Mathematics, Adelphi University, Garden City, L.I., New York 11530, USA

J. Kurzweil: Mathematical Institute of the Czechoslovach Academy of Science, Zitná 25, 115 67 Praha 1, Czechoslovachia

A. Källström: Department of Mathematics, University of Uppsala, Sysslomansgatan 8, 752 23 Uppsala, Sweden

I. Laine: Department of Mathematics, University of Joensuu, PL 111, SF-801 01 Joensuu 10, Finland

H. Langer: Sektion Mathematik, Technische Universität, DDR- 8027 Dresden, East Germany

A. Lasota: Institute of Mathematics, Silesian University, ul. Bankowa 14, 40-007 Katowice, Poland

S.J. Lee: Department of Mathematics, University of Alberta, Edmonton, Alberta, Canada

N.G. Lloyd: Department of Pure Mathematics, University College of Wales, Aberystwyth, Wales

M. Lopez: Department of Mathematics, McMaster University, Hamilton, Ontario L8S 4K1, Canada

L. Lorch: Department of Mathematics, York University, 4700 Keele Street, Downsview, Ontario M3J 1P3, Canada

W.A. Marcenko: Phys.-Techn. Institute of Low Temperatures, Lenin Prospect 47, Kharkov 310 086, USSR

J.B. McLeod: Mathematical Institute, 24-29 St. Giles, Oxford OX1 3LB, England

J. Moser: Courant Institute, 251 Mercer Street, New York, N.Y. 10012, USA

H.-D. Niessen: D 515 Bergheim-Zieverich, An der Bröelhecke 1, West Germany

I. Nåsell: Department of Mathematics, Royal Institute of Technology, S-100 44 Stockholm, Sweden

C. Olech: Instytut Matematyczny Polskiej Akademii Nauk, Sniadeckich 8, skr. poczt. 137, 00-950 Warszawa, Poland

C.G.C. Pitts: School of Mathematics and Physics, University of East Anglia, University Plain, Norwich NR4 7TJ, England

Å.V.C. Pleijel:	Department of Mathematics, University of Uppsala, Sysslomansgatan 8, 752 23 Uppsala, Sweden
S. Rankin:	Department of Mathematics, Murray State University, Murray, Kentucky 42071, USA
T.T. Read:	Department of Mathematics, Western Washington State College, Bellingham, Washington 98225, USA
F.S. Rofe-Beketov:	Physico-Technical Institute of Low Temperatures, Lenin Prospect, 47, Kharkov, 310086, USSR
K. Schmitt:	Department of Mathematics, University of Utah, Salt Lake City, Ut. 84112, USA
I.P. Stavroulakis:	Department of Mathematics, University of Ioannina, Ioannina, Greece
B. Szökefalvi-Nagy:	Bolyai Institute, Aradi vértanuk tere 1, 6720 Szeged, Hungary
P. Volkmann:	Mathematisches Institut, Universität Karlsruhe (TH), Postfach 6380, 75 Karlsruhe 1, West Germany
J. Vosmansky:	Department of Mathematical Analysis, University of J.E. Purkyne, Janáckovo Nám. 2a, 662 95 Brno, Czechoslovachia
J. Walter:	Institut für Mathematik, RWRH Aachen, Templergraben 55, D 51 Aachen, West Germany
J. Weidmann:	Fachbereich Mathematik der Universität Frankfurt, 6 Frankfurt am Main 1, Robert-Mayer-Strasse 10, West Germany
K.-O. Widman:	Department of Mathematics, University of Linköping, Fack, 581 83 Linköping, Sweden
A.D. Wood:	Mathematics Department, Cranfield Institute of Technology, Cranfield, Bedford MK 40 OAL, England
L.E. Zachrisson:	Department of Mathematics, Royal Institute of Technology, S - 100 44 Stockholm, Sweden
A. Zettl:	Department of Mathematical Sciences, Northern Illinois University, Dekalb, Illinois 60115, USA

DEFICIENCY-INDEX THEORY IN THE MULTI-PARAMETER STURM-LIOUVILLE CASE

F. V. Atkinson (Toronto)

1. Introduction.

The past fifteen years have seen the foundation of a systematic theory of eigenvalue problems involving several parameters. This has been carried out in the contexts of matrices, of symmetric operators, and of second-order difference and differential equations. The last area, that of multi-parameter Sturm-Liouville theory, may well be viewed as the origin of the whole topic of simultaneous eigenvalue problems, and has proved particularly fruitful to recent investigators. In this Sturm-Liouville setting, a natural approach for the analyst is to attempt to extend to the several-parameter case known results for the classical one-parameter theory.

The most important single result in classical Sturm-Liouville theory must surely be that of the completeness of the eigen-functions. Accordingly, and in view of its physical importance, it has been appropriate to concentrate efforts on establishing this same completeness in the multi-parameter case. Early partial results of Hilbert (13) and Dixon (9) have now been superseded by theories of a general character, developed by Browne, Faierman, Källström and Sleeman, and in part covering singular cases involving infinite intervals. For non-singular cases, the oscillatory characterisation of the eigenvalues, associated with the names of Klein and Richardson, may be considered as dealt with. Some more difficult problems for the non-singular case, such as estimates for eigenvalues and eigenfunctions and pointwise convergence of eigenfunction expansions have been considered by Faierman (12). A problem of mixed type was dealt with in (18). The rapidly expanding literature on the spectral theory of ordinary differential equations exhibits an almost infinite variety of topics and problems which one would wish to carry over to the multi-parameter context. We begin here by considering the classical limit-circle/limit-point classification of Weyl; the transposition of this to the multi-parameter case was initiated by Sleeman (19).

2. Classification for a single equation in several parameters.

For definiteness, we consider the case of a semi-infinite interval, and an equation

$$y''(x) + \{\sum_{1}^{k} \lambda_s p_s(x) - q(x)\} y(x) = 0 \ , \ a \leqslant x < \infty \ . \quad (2.1)$$

Here $p_s(x)$, $q(x)$ are to be continuous on the domain concerned, and real-valued. We then distinguish the following two main cases:

Case LC: Every solution of (2.1) satisfies

$$\int_a^\infty |p_s(x)y^2(x)|\,dx < \infty \quad , s = 1, \ldots, k , \qquad (2.2)$$

and Case LP: There is a solution of (2.1) for which one of (2.2) is false.

It will be noted that in this classification we have not specified any particular set of values of the parameters. In fact, standard arguments yield

Theorem 1. The above LC, LP classification is independent of the choice of the λ_s , $s = 1, \ldots, k$.

In other words, if all solutions of (2.1) satisfy (2.2) for some set of values of the parameters, real or complex, then this is so for all other sets of parameter-values; we did not even need the $p_s(x)$ to be real-valued.

A refinement of this classification deserves note. Taking the $p_s(x)$ to be real, we may for each set of real parameter-values ask whether (2.1) is oscillatory or not (i.e. whether every solution has an infinity of zeros). In the LC case, the answer does not depend on the choice of these real parameter-values. Accordingly, we may subdivide the LC case into LC-oscillatory and LC-non-oscillatory. This distinction is less apparent in the one-parameter case, that of

$$y''(x) + (\lambda p(x) - q(x))y = 0 , \quad x \geqslant a , \qquad (2.3)$$

since in the standard case $p(x) = 1$ the limit-circle case is well-known to be oscillatory.

It is also possible to subdivide these cases according to whether other square-integrable properties hold. Such procedures are given by Sleeman ((19) 210-211, 219-221), to which we refer for details. We use only the above classifications.

As noted by Sleeman (19) , the Weyl limit-point, limit-circle construction can be carried out for sets of parameter-values such that

$$\mathrm{Im}\ (\lambda_1 p_1(x) + \ldots + \lambda_k p_k(x)) > 0 , (< 0), \quad x \geqslant a ; \qquad (2.4)$$

there is then a non-trivial solution such that

$$\int_a^\infty |\mathrm{Im}\ (\lambda_1 p_1(x) + \ldots + \lambda_k p_k(x))|\,|y^2(x)|\,dx < \infty . \qquad (2.5)$$

The existence of such sets of λ_s is of course not automatic. However in cases of the most interest there is satisfied a rather stronger property, namely the following

Hypothesis: There is a linear combination $p^*(x)$ of the $p_s(x)$ such that

$$p^*(x) \geqslant |p_s(x)| , \quad s = 1, \ldots, k, \quad x \geqslant a . \tag{2.6}$$

There will then be sets of values of the Im λ_s such that (2.5) implies (2.2).

3. Deficiency-indices.

We may tentatively associate with (2.1) a pair of integers (in fact in this case the same) , following the pattern of the one-parameter case, as the upper bound for the dimension of the set of solutions of (2.1) satisfying (2.2), taken over the $(\lambda_1, \ldots, \lambda_k)$-sets defined by (2.4). More precisely, and assuming the above hypothesis, we define these integers as the dimension of the space of solutions satisfying (2.2), where for some $K > 0$ we have

$$\mathrm{Im}\ (\lambda_1 p_1(x) + \ldots + \lambda_k p_k(x)) \geqslant |Kp_s(x)| , \quad s = 1, \ldots, k, \ x \geqslant a; \tag{3.1}$$

the same number is, of course given, by taking the set of parameter-values for which the inequality in (3.1) is reversed.

Alternative interpretations of these numbers are available. It is natural to wish to express "deficiency-indices" as co-dimensions of the ranges of the operators concerned. While this can be done in the present case, it must be pointed out that the co-dimension is with respect to any one of a family of hilbert spaces. Let

$$g = f'' - qf + (\lambda_1 p_1 + \ldots + \lambda_k p_k)f , \quad f \in C_0(a, \infty) , \tag{3.2}$$

and let y be a non-trivial solution of (2.1) satisfying (2.5), where we suppose the parameters to satisfy (3.1). We then have

$$\int_a^\infty gy\ dx = 0 \quad . \tag{3.3}$$

This can be interpreted by saying that g/p^* is orthogonal to $\overline{y}$ in the weighted L^2-space over (a, ∞), with $p^*(x)$ as weight-function. Of course, $p^*(x)$ is not uniquely fixed by (2.6), though the various resulting L^2- norms will be topologically equivalent.

The equality of the "deficiency-indices" associated with (3.1), and the same set with reversed inequality, is liable to disappear if we go over to, say, odd order equations. This may be seen immediately by carrying through the same constructions for the first-order equation

$$y'(x) + i(\lambda_1 p_1(x) + \ldots + \lambda_k p_k(x) - q(x))y(x) = 0 , \quad x \geqslant a .$$

4. The simultaneous limit-circle case.

I pass now to the case of k simultaneous equations such as (2.1), or to

$$y_r''(x_r) + (\lambda_1 p_{r1}(x_r) + \ldots + \lambda_k p_{rk}(x_r) - q_r(x_r))y_r(x_r) = 0 ,$$

$$a_r \leqslant x_r < \infty \quad , \quad r = 1, \ldots, k . \tag{4.1}$$

As before, the $p_{rs}(x_r)$, $q_r(x_r)$ are real and continuous. To take a specific case, let us suppose that all k equations are in the LC-case, in the above classification; the "deficiency indices" will then be (2, 2) in each of the k cases. We pass over, for the moment, the question of whether they are oscillatory or not. We assume further that the "hypothesis" (2.6) is available in each case, and that the system is "right-definite", in that

$$\det p_{rs}(x_r) > 0 , \tag{4.2}$$

for all sets of values of the x_r .

The significance of the deficiency-indices is then that we obtain a well-defined eigenvalue problem, with expansion theorem, by associating two boundary conditions with each of (4.1). At the lower, non-singular end we may take a Sturmian one-point condition of the form

$$y_r(a_r)\cos \alpha_r = y_r'(a_r)\sin \alpha_r \ , \ r = 1, \ldots, k . \tag{4.3}$$

For boundary conditions at infinity, we choose a set $y_{ro} \not\equiv 0$ of solutions of (4.1), (4.3) for some selected set λ_{ro} , r = 1, ..., k, of real values of the parameters, and then impose the conditions

$$y_r(x_r)y_{ro}'(x_r) - y_r'(x_r)y_{ro}(x_r) \to 0 \text{ as } x_r \to \infty . \tag{4.4}$$

The eigenvalues of the problem are then k-tuples for which (4.1) are (4.3-4) have non-trivial solutions $y_r(x_r)$, r = 1, ..., k. The product of these functions is, as usual, the eigenfunction associated with this eigenvalue.

In justification we note that the LC hypothesis ensures the existence of the limit in (4.4). Reality of the eigenvalucs may then be deduced from (4.2), and likewise orthogonality of eigenfunctions and discreteness of the spectrum. The completeness of the eigenfunctions may be deduced from that for the finite-interval case. One imposes (4.4) with equality when $x_r = b_r < \infty$, and then makes the b_r tend to infinity; it is not necessary to use the Helly selection theorem.

5. Characterization of eigenvalues in the full limit-circle case.

Once more we may look to the one-parameter case, with a view to the extension of results to the multi-parameter case. We consider the rate of growth of eigenvalues in the limit circle case, and their oscillatory characterisation.

On the first aspect, one one may note the approach which exhibits the eigenvalues as the zeros of an entire function, for the one-parameter case. The rate of growth of the eigenvalues can then be related to the order of the entire function whose zeros are the eigenvalues. This approach can be used both in the non-singular finite-interval case, and also in the "quasi-singular case", when the interval is infinite but the limit-circle case holds.

In the multi-parameter case this approach proves difficult, but is worth pursuing for some distance. We denote by y_r, y_{ro} as before solutions of (4.1) with parameter-values $\lambda_1, \dots$ and $\lambda_{1o}, \dots$ and initial data $y_r(a_r) = \sin \alpha_r$, $y_r'(a_r) = \cos \alpha_r$, and likewise for y_{ro}. We use similarly z_r, z_{ro} with the initial data replaced by $\cos \alpha_r$, $-\sin \alpha_r$. We then define

$$u_r = z_{ro}y_r' - z_{ro}'y_r , \quad v_r = y_r y_{ro}' - y_r' y_{ro} . \qquad (5.1\text{-}2)$$

Some easy calculations give

$$u_r' = z_{ro}(u_r y_{ro} + v_r z_{ro}) \sum (\lambda_{so} - \lambda_s) p_{rs} , \qquad (5.3)$$

$$v_r' = - y_{ro}(u_r y_{ro} + v_r z_{ro}) \sum (\lambda_{so} - \lambda_s) p_{rs} . \qquad (5.4)$$

These differential equations and the classical Weyl reasoning show, in the limit-circle case, that $u_r(x_r)$, $v_r(x_r)$ tend (as already noted) to limits as $x_r \to \infty$, which are entire functions $U_r(\lambda)$, $V_r(\lambda)$ of the parameters $\lambda_1, \dots, \lambda_k$. In particular, the eigenvalues are the simultaneous zeros of the k functions $U_r(\lambda)$. However, it does not seem that in the multi-parameter case much can be deduced from this fact, even with information on the order of these entire functions.

As sometimes happens in the multi-parameter theory, progress can be made with more primitive arguments. We introduce the Prüfer-type substitution $\tan \theta(x_r) = v_r(x_r)/u_r(x_r)$, and find that

$$\theta_r' = (y_{ro} \cos \theta_r + z_{ro} \sin \theta_r)^2 \sum (\lambda_{so} - \lambda_s) p_{rs} . \qquad (5.5)$$

Still in the limit-circle case, we see that $\theta_r(x_r)$ tends as x_r tends to infinity to a limit $\Theta_r(\lambda)$, and that eigenvalues are characterised by the requirements

$$\Theta_r(\lambda) = n_r \pi , \quad r = 1, \ldots, k . \tag{5.6}$$

Here the n_r, $r = 1, \ldots, k$, are integers.

We approach an oscillatory characterisation of eigenvalues which, for the multi-parameter case, generalises the Klein oscillation theorem for the finite-interval case (which in turn extends the classical Sturm oscillation theorem), and on the other hand extends little known oscillatory properties of the one-parameter limit-circle case.

Theorem 2. With the assumptions of Section 4, and for a given selection of the base eigenvalue $\lambda_{1o}, \ldots, \lambda_{ko}$, the general eigenvalue $\lambda_1, \ldots, \lambda_k$ is uniquely determined by the integers $n_1, \ldots, n_k$. There holds the relation

$$\sum |n_r| = o\{ \sum |\lambda_s| \} . \tag{5.7}$$

That there can correspond at most one eigenvalue λ to any given set of integers n_r in (5.6) follows from (5.5) together with the fact, a consequence of (4.2), that if λ, λ' are distinct real k-tuples, then for some r the expression

$$\sum (\lambda_s - \lambda_s') p_{rs}(x_r) ,$$

does not change sign, and does not vanish identically.

The asymptotic relationship (5.7), weakened to "big O" on the right, follows by an integration of (5.5), together with the limit-circle hypothesis. The proof of the full result involves applying (5.5) over some interval (b_r, ∞), and a distinct argument over (a_r, b_r).

In Theorem 2, we have not determined the admissible range of values of the sets of integers $n_1, \ldots, n_k$. It seems clear that this will depend on whether any or all of (4.1) are oscillatory.

In the one-parameter case, (5.7) is related to the question of the order of the entire function whose vanishing determines the eigenvalues in the limit circle case (or which appears in the denominator of the function $m(\lambda)$); this question has been raised by Everitt. Closely connected with this is the question of the exponent of convergence of the eigenvalues, a topic recently investigated by Weidmann (22), in a more general setting. These questions are being investigated currently by Fulton and the author.

6. Spectral functions, $m(\lambda)$ and the Green's function.

I turn now to a rather more problematic area. The functions just named are not only important but also are intimately related in the one-parameter theory. Their extensions to the multi-parameter case exist, and have great importance, though their inter-relations are not so clear.

For the finite-interval case one introduces the idea of a spectral function by means of the formulae

$$g(\lambda) = \int f(x)y(x,\lambda)p(x)dx , \quad f(x) = \int g(\lambda)y(x,\lambda)d\tau(\lambda) , \tag{6.1}$$

together with the Parseval equality

$$\int |f|^2 p\, dx = \int |g|^2 d\tau ; \tag{6.2}$$

the standard theory of (2.3), for a finite interval (a, b) and with the usual one-point or separated boundary conditions, exhibits $\tau(\lambda)$ as a step-function with jumps at an eigenvalue λ of amount

$$\{\int |y(x.\lambda)|^2 p(x)\, dx\}^{-1} .. \tag{6.3}$$

Here f is any function in L^2 (with weight-function p), and y is a solution of (2.3) with initial data fixed as previously and satisfying the initial boundary condition. A spectral function for the interval (a, ∞) may be obtained by making $b \to \infty$. and using Helly's selection and integration theorems, as in (8,233-234).

All this extends to the multi-parameter case, granted the completeness of the eigenfunctions in the finite-interval case. In (6.1), x must stand for $x_1, \ldots, x_k$, dx for $dx_1 \ldots dx_k$, $p(x)$ for $\det p_{rs}(x_r)$, λ for $\lambda_1, \ldots, \lambda_k$, $y(x, \lambda)$ for $\Pi y_r(x_r, \lambda)$, and $d\tau(\lambda)$ is suitably interpreted as a mass-distribution in R^k ; for example, in the finite-interval case, (6.3) gives the point-mass at an eigenvalue. The extension to the singular case by way of a limiting process and selection theorems has been carried out by Browne (6) It would seem that little is known about the nature, as regards continuity or rate of growth, of these multi-dimensional spectral functions.

In the one-parameter case, a more sophisticated theory has been developed by M. G. Krein and his school (see e.g.(15)), in which a spectral function is defined, roughly speaking, as a non-decreasing function satisfying (6.1-2), and the problem is posed of determining all such functions; in particular, the subclass of orthogonal spectral functions is to be determined, these being such that the maps (6.1-2) onto the appropriate hilbert spaces. These notions await investigation in the multi-parameter case.

For the one-parameter finite-interval case, certain (non-orthogonal spectral functions can be given explicitly, that is to say without reference to eigenvalues ((1), Section 8.11); in these, $\tau(\lambda) \in C'(-\infty,$ It is perhaps noteworthy that this can also be done in our case. Assuming (4.2), we have, with the usual Sturmian boundary conditions,

Theorem 3. For the multi-parameter finite-interval case, there is a spectral function for which $d\tau(\lambda)$ is to be replaced by

$$\pi^{-k} \Pi \{c_r y_r^{2}(b_r, \lambda) + c_r^{-1} y_r'^{2}(b_r, \lambda)\}^{-1} d\lambda ; \qquad (6.$$

here $c_1, \ldots, c_k$ are any positive numbers and $d\lambda = \Pi d\lambda_r$.

The proof proceeds by integration over the boundary condition angles β_r at the upper ends b_r of the basic intervals, just as in the one-parameter case.

Concerning the one-parameter case, see also the remarks in

Expressions of the type (6.4) can be used to discuss the nature of the spectrum in singular cases, by making the $b_r \to \infty$. In doing this we aim to choose the c_r , possibly dependent on the b_r , so that (6. tends to a limit.

We turn to aspects involving $m(\lambda)$, or its analogues. We are, in the one-parameter case concerned with functions

$$m(\lambda, b, \beta) = -(z(b, \lambda)\cos\beta - z'(b, \lambda)\sin\beta)/(y(b, \lambda)\cos\beta - y'(b, \lambda)\sin\beta),$$

where y, z are solutions of (2.3) with the initial data $y = \sin\alpha$, $y' = \cos\alpha$, $z = \cos\alpha$, $z' = -\sin\alpha$, and with the limiting behaviour of this expression as $b \to \infty$, possibly with varying β . Salient facts include the following:

(i) if $b < \infty$, m is a meromorphic function of λ , with poles at t eigenvalues given by taking α, β as boundary-condition angles,

(ii) m is a function of Pick-Nevanlinna type, having imaginary part of fixed sign in the upper and lower half-planes (14),

(iii) m can be expressed as an integral over the real axis, togethe with a linear term, the integral involving the spectral function,

(iv) the spectral function may be expressed as the limit of an integ involving m ,

(v) $m(\lambda, b, \beta)$ describes, for varying β, for fixed non-real λ and fixed $b > a$, a circle which shrinks, possibly to a point, as $b \to \infty$,

(vi) the Green's function of the problem (with $b < \infty$) involves m

Regarding (iii) for the limit-point case, and the closely related pproperty (iv) (the Titchmarsh-Kodaira formula) see (8 , Chapter 9).

In the multi-parameter case (4.1), it is natural to form k functions m_r, of $\lambda_1, \ldots, \lambda_k$, b_r and $\beta_1, \ldots, \beta_k$. The nesting-circle aspect has, as noted above, been discussed by Sleeman (19), along with the problem of the significance of the singularities of these functions.

I am indebted to Professor Everitt for calling attention to the Pick-Nevanlinna aspect (ii). The sign of Im m_r is related to the sign of the left of (2.4), when that is fixed. One can thus say that if the definiteness condition (4.2) holds, then at least one of the m_r has non-zero imaginary part, if the same is true for the λ_r. However, this does not seem to relations of the form (iii), (iv).

One may conjecture that, in place of (iii), one should consider representations of the product $m_1 \ldots m_k$, or again the product of Green's functions for the separate intervals. This would form a distinct topic from the use of Green's functions for partial differential equations associated with (4.1) in the multi-parameter case, which is a known successful method for handling the problem of completeness of eigenfunctions but which involves the difficulty of establishing the existence of this function, to say nothing of expressing it.

References

1. Atkinson, F.V.: "Discrete and continuous boundary problems", (Academic Press, New York, 1964).
2. Atkinson, F. V.: Boundary problems leading to orthogonal polynomials in several variables, Bull. Amer. Math. Soc. 69(1963), 345-351.
3. Atkinson, F. V.; Multiparameter spectral theory, Bull. Amer. Math. Soc. 74(1968), 1-27.
4. Atkinson, F.V.: "Multiparameter eigenvalue problems", Vol. I, (Academic Press, New York, 1972).
5. Browne, P.J.: A multi-paramer eigenvalue problem, Jour. Math. Anal. Appl. 38(1972), 553-368.
6. Browne, P.J.: A singular multi-parameter eigenvalue problem in second order ordinary differential equations, Jour. Diff. Equ. 12(1972), 81-94.
7. Browne, P. J.:Multi-parameter spectral theory, Indiana Univ. Math. J., 24(1974), 249-257,
8. Coddington, E.A. and Levinson, N.: "Theory of ordinary differential equatinns", (McGraw-Hill, New York, 1955).
9. Dixon, A.C.:Harmonic expansions of functions of two variables, Proc. London Math. Soc.(2)5(1907), 411-478.
10. Faierman, M.:The completeness and expansion theorems associated with the multi-parameter eigenvalue problem in ordinary differential equations, Jour. Diff. Equ. 5(1969), 197-213.

11. Faierman, M.: Asymptotic formulae for the eigenvalues of a two-pa meter ordinary differential equation of the second order, Trans. Amer. Math. Soc. 168(1972), 1-52.

12. Faierman, M.: Asymptotic formulae for the eigenvalues of a two-parameter system of ordinary differential equations of the second order, Canad. Math. Bull. 17(1975), 657-665.

13. Hilbert, D.:"Grundzüge einer allgemeinen Theorie der linearen Integralgleichungen" , (Teubner, Leipzig, 1912).

14. Kac, I.S. and Krein, M.G.: R-functions - analytic functions mapping the upper halfplane into itself, Amer. Math. Soc. Transl. (2), 103(1974), 1-18 (= Supplement to Russian edition of reference 1 , (Moscow, 1968)).

15. Kac, I.S. and Krein, M.G.: On the spectral functions of the stri Amer. Math. Soc. Transl.(2), 103(1974), 19-102 (= Supplement II Russian edition of reference 1).

16, Källström, A. and Sleeman, B.D.: A left definite multiparameter eigenvalue problem in ordinary differential equations, Proc. Roy Soc. Edin. 74A, 11(1974/75), 145-155.

17. Källström, A. and Sleeman, B.D.: An abstract multiparameter eige value problem, Uppsala University Mathematics Report No.1975:2.

18. Ma, S.P.: "boundary value problems for a matrix-differential sys tem in two parameters", (Thesis, University of Toronto, 1972).

19. Sleeman, B.D.: Singular linear differential **operators** with many parameters, Proc. Roy. Soc. Edin.(A), 17(1972/73), 199-232.

20. Sleeman, B.D.: Completeness and expansion theorems for a two-parameter eigenvalue problem in ordinary differential equations using variational principles, Jour. London Math. Soc.(2), 6(1973), 705-712.

21. Titchmarsh, E.C.: "Eigenfunction expansions associated with second-order ordinary differential equations", Part I, 2nd edn., (Oxford University Press, 1962), Part II (Oxford University Press, 1958).

22. Weidmann, J.: Verteilung der Eigenwerte für eine Klasse von Integraloperatoren in $L_2(a, b)$, Jour. für Mathematik, 276(1974), 213-220.

ALTERNATE METHOD, FINITE ELEMENTS, AND ANALYSIS IN THE LARGE

Lamberto Cesari

Introduction. In this lecture we present some of the existence theorems in the large which have been recently obtained by a priori bounds, topological arguments, and the alternative method: theorems concerning forced oscillations (harmonics) of Liénard systems under sole qualitative hypotheses (§2), and theorems of the Landesman and Lazer type (§3) which are presented here in the form of existence theorems "across a point of resonance". In the case of forced oscillations these theorems correspond to the well known phenomenon of entrainement of frequency. In §§5,6,7 we present results which have ensued by the association of the alternative method and finite elements theory. In particular, a method of successive approximations is presented here for nonlinear boundary value problems whose underlying linear operator is not necessarily self-adjoint. Sufficient conditions for its convergence are given.

1. The alternative, or bifurcation process. We present here first the basic framework of the alternative, or bifurcation method [6] as described in more details in recent expositions [7,8].

Let X,Y be Banach spaces, let $E : \mathcal{D}(E) \to Y$, $\mathcal{D}(E) \subset X$, be a linear operator with domain $\mathcal{D}(E)$, and let $N : \mathcal{D}(N) \to Y$ be a continuous nonnecessarily linear operator, $\mathcal{D}(E) \cap \mathcal{D}(N) \neq \emptyset$. Let $X = X_o + X_1$, $Y = Y_o + Y_1$ be decompositions of X and Y, and $P : X \to X$, $Q : Y \to Y$ projection operators (i.e., linear, bounded, and idempotent), such that $X_o = PX$, $X_1 = (I-P)X$, $Y_o = QY$, $Y_1 = (I-Q)Y$, and assume that the null space of E is contained in X_o and the image of $X_1 \cap \mathcal{D}(E)$ is Y_1, or $\ker E \subset X_o$ and $Y_1 = E(X_1 \cap \mathcal{D}(E))$. Then, the map $E : X_1 \cap \mathcal{D}(E) \to Y_1$ is one-one and onto, and the partial inverse $H : Y_1 \to X_1 \cap \mathcal{D}(E)$ exists as a linear map. In [7,8] we have discussed these assumptions together with the natural relations: (k_1) $H(I-Q)E = I-P$, (k_2) $QE = EP$, (k_3) $EH(I-Q) = I-Q$. As proved in [7,8], the operator H is bounded if both the graph and the range of E are closed. Also, equation

$$Ex = Nx \tag{1}$$

is equivalent to the system of auxiliary and bifurcation equations

$$x = Px + H(I-Q)Nx\ , \tag{2}$$

$$Q(Ex - Nx) = 0\ . \tag{3}$$

If $x^* = Px$ and T denotes the map defined by $Tx = x^* + H(I-Q)Nx$, then the auxiliary equation takes the form of a fixed point problem $x = Tx$. For $y = x - Px$, the auxiliary equation takes the form of a Hammerstein equation $(I-K\overline{N})y = 0$, where

$K = H(I-Q)$, and $\overline{N}y = N(x^*+y)$.

For every $x^* \in X_o$, we have $PTx = x^* = Px$, that is, T maps the fiber $P^{-1}x^*$ of P through x^* into itself. Thus, if for every x^* the transformation $T : P^{-1}x^* \to P^{-1}x^*$ is a contraction, then it has a unique fixed point $x = Tx = \mathcal{T}x^*$ which depends on x^* . The bifurcation equation then becomes $Q(E-N)\mathcal{T}x^* = 0$, an equation in X_o (See [7], [8] for applications).

If in (1), Nx is replaced by εNx , ε a small parameter and N is locally Lipschitzian, then T is a local contraction for $|\varepsilon|$ sufficiently small. This al occurs in problems of nonlinear eigenvalues, or bifurcation theory proper. (See for these applications, e.g., Hale and Gambill [28], Hale [27, particularly pp. 34-42], Nagle [35], Bowman [5]). For T a contraction, applications have been made to the determination of an error bound ρ , or $\|x-x_o\| \le \rho$, for the difference between a known (Galerkin) approximation x_o and the exact solution x (see Cesari and Bowman [16]).

For the original equation (1)(large nonlinearities) with N locally Lipschitzi it was actually shown in [6,7,8] for self-adjoint problems, and in [8,27] for a larg class of nonself-adjoint problems, that it is always possible to choose the space $X_o \supset \ker E$ and the operators P,Q,H in such a way to make T a local contraction. The same result will be obtained here in §5 for general nonself-adjoint problems by the use of finite elements. A modified alternative scheme for general nonself-adjoi problems for which also T can always be made into a contraction, has been recently studied by McKenna [34].

For a treatment of auxiliary and bifurcation equations in the case in which $-E$ and N are monotone maps we refer to [8] and [17].

In general, it may be convenient to discuss equations (2) and (3) as a system. For instance, for $X_o = \ker E$, the bifurcation equation reduces to $QNx = 0$, and the solutions of system (2), (3) can be thought of as the fixed points of the transforma- tion $T^* : (x,x^*) \to (\overline{x},\overline{x}^*)$ defined by

$$\overline{x} = x^* + H(I-Q)Nx \ , \ \overline{x}^* = x^* + QNx \ , \tag{4}$$

or analogously, for $y = x - Px$, as the fixed points of the transformation $T_1^* : (y,x^*) \to (\overline{y},\overline{x}^*)$ defined by

$$\overline{y} = H(I-Q)N(x^*+y) \ , \ \overline{x}^* = x^* + QNx \ ,$$

or of analogous transformations. Also, if $S : Y_o \to X_o$ is any continuous map with $S^{-1}(0) = 0$, then the solutions of system (2),(3) are also the solutions of the sole equation in the space X:

$$T^{**}(x) = y - H(I-Q)N(x^*+y) + SQN(x^*+y) = 0 \ ,$$

where $x = x^* + y$, $x^* \in X_o$, $y \in X_1$, an equation proposed by Williams [39] in 1968, and further studied by Mawhin after 1972 (coincidence degree). The essential equivalence of these approaches is confirmed by studies showing that the various maps above have the same topological degree (see, e.g., Williams [39]. More work toward this equivalence is in progress.).

2. Forced oscillations of Liénard systems. The results of this section have just been obtained by the use of transformations similar to (4), by establishing a priori bounds, and by the use of the Ulam-Borsuk topological lemma.

Let us consider a Liénard system with forcing terms of the type

(5) $$x''(t) + (d/dt)F(x(t)) + (d/dt)V(x(t),t) + Ax + g(x(t)) = e(t) ,$$

where $x(t) = (x_1,\ldots,x_n)$, $-\infty < t < +\infty$, $A = [a_{ij}]$, $F(x) = (F_1,\ldots,F_n) = \operatorname{div} G(x)$, $V(x,t) = (V_1,\ldots,V_n)$, $g(x) = (g_1,\ldots,g_n)$, $e(t) = (e_1,\ldots,e_n)$, where we assume that $e(t)$ is a 2π-periodic continuous function of mean value zero, that A is a constant $n \times n$ matrix, that $G(x)$ is of class C^2 in R^n, that $V(x,t)$ is 2π-periodic in t for every x and is such that all V_i, $\partial V_i/\partial x_j$ are of class C^1 in R^{n+1}, and that $g(x)$ is continuous in R^n with $g(x)/|x| \to 0$ as $|x| \to \infty$.

(2.i) Theorem (Cesari, Kannan, DeVries). Under the hypotheses above system (5) has at least one 2π-periodic solution $x(t)$, $-\infty < t < +\infty$, provided that one of the following assumptions hold. Either

(α) there are nonnegative constants c,d,C,D,C',D', $c > 0$, and integer $p \geq 2$ such that $xF(x) = \Sigma_1^n x_iF_i(x) \geq c|x|^{2p} + d$, $V(x,t) = V_1(x,t) + V_2(x,t)$, $V_1(x,t) = S(t)$ grad $W(x)$ for some C^2 scalar function $W(x)$ and some 2π-periodic C^1 function $S(t) = (S_1,\ldots,S_n)$, $|V_1(x,t)| \leq C'|x|^{2p-2} + D'$, $|V_2(x,t)| \leq C|x|^p + D$, and moreover $\det A \neq 0$.;

or (α') $g \equiv 0$, F and V as in (α), A an arbitrary constant matrix.;

or (β) the same as in (α) with $p = 1$ and $c > C + \|A-A_{-1}\|/2$;

or (β') $g \equiv 0$, F and V as in (β), A an arbitrary constant matrix;

or (γ) the same as in (α) with $|V_1(x,t)| \leq C'|x|^{2p-1} + D'$, $c > C'$ for $p > 1$, and $c > C + C' + \|A-A_{-1}\|/2$ for $p = 1$;

or (γ') $g \equiv 0$, F and V as in (γ), A an arbitrary constant matrix.

This theorem was proved by Cesari and Kannan [18,19] for the case $V_1 \equiv 0$, G homogeneous of degree $2p$ and constant sign, by alternative method, a priori bounds, and the Borsuk-Ulam lemma. By the same argument DeVries [24] extended the theorem to the form stated above. The present form was suggested by the requirement to make it invariant with respect to translations $x = y + \phi(t)$, ϕ 2π-periodic and smooth.

For further results concerning systems analogous to (5) of the Liénard or Rayleigh types under conditions of symmetry, we refer to Bowman [4] and DeVries [24], where use is made of Hale's general concept of symmetry.

3. Theorems of the Landesman-Lazer type. The results below are contained in work by Cesari [8,13,14,15] and by Cesari and Kannan [20], where a unification and extension w[as] obtained of previous specific theorems at resonance, each with specific and rather different assumptions (Lazer and Leach [33], Landesman and Lazer [32], Williams [40], DeFigueiredo [23], Necas [36], Hess [29], Fucik [26], Chang [22], and others). It was shown indeed by Cesari and Kannan that a suitable abstract unilateral relation suffices for the existence. The proofs are based on a priori bounds, alternative method, and Schauder's fixed point theorem for the map T^*, in the same line therefore of the proofs of Landesman and Lazer, and of Williams in their specific theorems. However, as mentioned at the end of §2, analogous proofs could be given in terms of Leray-Schauder argument and topological degree for the maps T^*, or T_1^*, or T^{**}.

Let X and Y be real Banach spaces, and P,Q,H operators as in §1, but now we assume that N is defined in the whole of X, and that Y is a space of linear operators on X so that an operation $\langle y,x\rangle$ is defined from $X \times Y$ into the reals, is linear both in x and y, and has the following properties: (π_1) $|\langle y,x\rangle| \leq K\|x\|\,\|y\|$ for all $x \in X$, $y \in Y$, and some constant $K > 0$; (π_2). For $y \in Y_o$ we have $y = 0$ if and only if $\langle y,x^*\rangle = 0$ for all $x^* \in X_o$. It is always possible to choose the norms in X and Y, or to choose the operation $\langle y,x\rangle$, in such a way that $K = 1$. If $X = Y = S$, a real Hilbert space, then we can take for $\langle y,x\rangle$ the inner product in S and then $K = 1$. Let $L = \|H\|$, $k_o = \|P\|$, $k' = \|I-P\|$, $X = \|Q\|$, $X' = \|I-Q\|$.

Also, we assume that $X_o = PX = \ker E$ is of finite dimension m and not trivial thus, $1 \leq m < \infty$. Let $w = (w_1,\dots,w_m)$ be a basis in $X_o = \ker E$. By $\langle y,w\rangle$ we denote the m-vector $(\langle y,w_i\rangle,\ i = 1,\dots,m)$. For every $x^* \in X_o$ we have $x^* = \Sigma_1^m c_i w_i$, or briefly $x^* = cw$, $c = (c_1,\dots,c_m) \in R^m$, and there are constants $0 < \gamma' \leq \gamma < \infty$ such that $\gamma'|c| \leq \|cw\| \leq \gamma|c|$ for all $x^* = cw \in X_o$ and where $|\ |$ denotes the Euclidean norm in R^m. Take $\gamma_o = \min[1,\gamma]$. Moreover, there is a constant $\mu > 0$ such that for every $y \in Y$ and $d = \langle y,w\rangle$, or $d = (d_1,\dots,d_m) \in R^m$, $d_i = \langle y,w_i\rangle$, $i = 1,\dots,m$, we have $|d| \leq \mu\|y\|$. Finally, we assume that H is a (linear bounded) compact operator. Below, α will be a real parameter.

Furthermore, we shall denote by A an arbitrary continuous operator $A : X \to Y$, not necessarily linear, but bounded, that is, mapping bounded subsets of X into bounded subsets of Y, or equivalently, satisfying a relation $\|Ax\| \leq \omega(\|x\|)$ for all $x \in X$ and some monotone nondecreasing function $\omega(\zeta) \geq 0$, $0 \leq \zeta < +\infty$, (ω not necessarily continuous, nor zero at the origin).

(3.i) Theorem (Cesari [14]). Under the assumptions above, if (B_o) there is a constan[t] $J_o > 0$ such that $\|Nx\| \leq J_o$ for all $x \in X$; and if (N_ε) there are constants $R_o \geq 0$

$\varepsilon > 0$, $K > LX'J_o$ such that $\langle QNx, x^* \rangle \leq -\varepsilon \|x^*\|$ [or $\langle QNx, x^* \rangle \geq \varepsilon \|x^*\|$] for all $x \in X$, $x^* \in X_o$ with $Px = x^*$, $\|x^*\| \geq R_o$, $\|x - x^*\| \leq K$, then there are also constants $\alpha_o > 0$, $C > 0$ such that, for every real α with $|\alpha| \leq \alpha_o$, equation $Ex + \alpha Ax = Nx$ has at least a solution $x \in \mathcal{D}(E) \subset X$ with $\|x\| \leq C$.

Since $\ker E$ is not trivial, $\alpha = 0$ is a point of resonance. Thus, as α crosses the point of resonance $\alpha = 0$, the solutions, of which the existence is stated in this theorem, are equibounded. For $\alpha = 0$, $\varepsilon = 0$, Theorem (3.i) reduces to a theorem at resonance. An analogous theorem holds for N of limited growth as $\|x\| \to \infty$:

(3.ii) Theorem (Cesari [14]). Under the same general assumptions of Theorem (3.i), if (B_k) there are constants $J_o \geq 0$, $J_1 > 0$, $0 \leq k < 1$, such that $\|Nx\| \leq J_o + J_1 \|x\|^k$ for all $x \in X$, and if $(N_{\varepsilon k})$ there are constants $R_o \geq 0$, $\varepsilon > 0$, $K_o > LX'J_o$ $K_1 > LX'J_1\gamma'^{-1}(k_o\gamma_o)^k$ such that $\langle QNx, x^* \rangle \leq -\varepsilon \|x^*\|^{1+k}$ [or always $\langle QNx, x^* \rangle \geq \varepsilon \|x^*\|^{1+k}$ for all $x \in X$, $x^* \in X_o$ with $Px = x^*$, $\|x^*\| \geq R_o$ $\|x - x^*\| \leq K_o + K_1\|x\|^k$; then the same conclusion of Theorem (3.i) holds.

Both theorems above are actually particular cases of the following statement in which no particular growth requirement is made, but it is requested that for some single fixed number S a certain inequality holds. To state the theorem we denote by R_o the constant which will appear in assumption (N_ϕ) below, and we choose arbitrary constants $\sigma_1, \sigma_2, \sigma, \lambda_o, \lambda_1$ with $0 < \sigma_1 < \sigma_2 < \sigma < \min[1, \gamma^{-1}]$, $\lambda_o \geq \max [1, \gamma'^{-1}k_o]$, $\lambda_1 < \min[(LX')^{-1}(1 - \gamma\sigma), (\mu X)^{-1}(\sigma - \sigma_2)]$.

(3.iii) Theorem (Cesari [14]). Under the same assumptions of Theorem (3.i), let $\phi(\zeta)$, $\phi_1(\zeta)$, $\psi(\zeta) \geq 0$, $0 \leq \zeta < +\infty$, be monotone nondecreasing functions, both ϕ_1 and ψ positive for $\zeta \geq R_o$. Let us assume that (B_ϕ) $\|Nx\| \leq \phi(\|x\|)$ for all $x \in X$; and that (N_ϕ) $\langle QNx, x^* \rangle \leq -\phi_1(\|x^*\|)$ [or $\langle QNx, x^* \rangle \geq \phi_1(\|x^*\|)$] for all $x \in X$, $X^* \in X_o$ with $Px = x^*$, $\|x^*\| \geq R_o$, $\|x - x^*\| \leq \psi(\|x\|)$. Let us assume further that there is a constant $S \geq \sigma_1^{-1}\lambda_o R_o$ with $\phi(S)/S < \lambda_1$ and $LX'\phi(S) < \psi(k_o^{-1}\gamma'\sigma_1 S)$. Then the same conclusion of Theorem (3.i) holds.

In Theorem (3.iii) we may also assume that (N_ϕ) holds for $\|x - x^*\| \leq \psi(\|x^*\|)$ and that in the last requirement we assume $LX'\phi(S) < \psi(\gamma'\sigma_1 S)$. Concerning the actual derivation of Theorems (i,ii) from (iii), and the actual derivation of the specific existence theorems of Lazer and Leach, Landesman and Lazer, Williams, etc., we refer to (Cesari [8,13,14,15]).

4. <u>A further decomposition of the spaces X and Y</u>. As in §1 let X, Y be Banach spaces, let $E : \mathcal{D}(E) \to Y$, $\mathcal{D}(E) \subset X$, be a linear operator, and $N : X \to Y$ a non-necessarily linear operator. Let $X_{oo} = \ker E$, $Y_{o1} = \mathcal{R}(E)$, with the assumption that

there are decompositions $X = X_{oo} + X_{ol}$, $Y = Y_{oo} + Y_{ol}$ and projection operators $P_o : X \to X$, $Q_o : Y \to Y$ such that $X_{oo} = P_oX$, $X_{ol} = (I-P_o)X$, $Y_{oo} = Q_oY$, $Y_{ol} = (I-Q_o)$ Then the map $E : X_{ol} \cap \mathcal{D}(E) \to Y_{ol}$ is one-one and onto, and we denote by H_o the partial inverse $H_o : Y_{ol} \to X_{ol} \cap \mathcal{D}(E)$ as we did in §1. Now the relations hold: $Q_oE = 0$, $EP_o = 0$, $E(I-P_o) = E$, $H_o(I-Q_o)E = H_oE = H_oE(I-P_o)$, in particular, relation (k_{123}) of §1, and equation $Ex = Nx$ in X is equivalent to the system of auxiliar and bifurcation equations

(6) $$x = P_ox + H_o(I-Q_o)Nx , \qquad Q_oNx = 0 .$$

For the next further decomposition of the spaces X,Y, the following lemma is helpful.

(4.i) Let $S : X \to X$ be any projection operator with $SX \subset X_{ol} \cap \mathcal{D}(E)$, $SP_o = 0$, $P_oS = 0$. Then, $P = P_o + S$ is also a projection operator, and system (6) is equiva lent to $x = Px + (I-P)H_o(I-Q_o)Nx$, $Sx = SH_o(I-Q_o)Nx$, $Q_oNx = 0$.

A proof of this lemma is essentially given in [8, p. 25].

Now let Σ' denote a closed subspace of Y_{ol} and let Σ be the corresponding subspace $\Sigma = H_o\Sigma' \subset X_{ol}$. Let $R' : Y_{ol} \to Y_{ol}$ be a projection operator with range Σ', or $R'(Y_{ol}) = \Sigma'$, and let us define the projection operator $S' : X_{ol} \to X_{ol}$ b taking $S' = H_oR'E$. Then, the range of S' is Σ, and for any $\bar{y} \in Y_{ol}$ we also have $R'\bar{y} = ES'H_o\bar{y}$. Now we take $S = S'(I-P_o)$, $R = R'(I-Q_o)$, so that $S : X \to X$, $R : Y \to Y$ are projection operators in X and Y respectively. Moreover, $P_oS = 0$, $SP_o = 0$, $Q_oR = 0$, $RQ_o = 0$.

(4.ii) Under the assumptions above, then $P = P_o + S$ is a projection operator in X $Q = Q_o + R$ is a projection operator in Y, and, for $X_o = PX$, $X_1 = (I-P)X$, $Y_o = QY$, $Y_1 = (I-Q)Y$, we have $X_o \supset \ker E = X_{oo}$, $Y_1 \subset \mathcal{R}(E) = Y_{ol}$. Moreover, if H denote the restriction of H_o to Y_1, or $H = H_o|_{Y_1}$, then the relations hold (k_1) $H(I-Q$ $I-P$, (k_2) $QE = EP$, (k_3) $EH(I-Q) = I-Q$. Finally, $I-Q = I-Q_o - R'(I-Q_o) = (I-R')(I-$

A proof of this statement is essentially given in [8, pp. 27-28]. Thus, we hav the final decompositions

$$X = X_o + X_1 = X_{oo} + \Sigma + X_1 , \quad Y = Y_o + Y_1 = Y_{oo} + \Sigma' + Y_1 ,$$

and with the operators P,Q,H, the equation $Ex = Nx$ is now equivalent to the syst of auxiliary and bifurcation equations

(7) $$x = Px + H(I-Q)Nx , \qquad Q(Ex - Nx) = 0 ,$$

though now we may write the auxiliary equation also in the form

(8) $$x = Px + H(I-R')(I-Q_o)Nx \ .$$

First, a few general considerations on relations (7) and (8). We shall think here that the image $R(N)$ of N is contained in a Banach space $Z \subset Y$, or $N : X \to Z$, $R(N) = NX \subset Z \subset Y$. We denote by $\| \ \|_X$, $\| \ \|_Y$, $\| \ \|_Z$ the norms in X,Y,Z, though we may omit the subscripts when evident from the context.

We shall think of the operators appearing in the auxiliary equation (8) as follows: $N : X \to Z$, $I-Q_o : Z \to Z$, $I-R' : Z \to Y$, $H_o : Y_{o1} \to X$, $P : X \to X$. Accordingly, we shall introduce corresponding constants as follows:

(9)
$$\| Nx_1 - Nx_2 \|_Z \le L \| x_1 - x_2 \|_X \quad \text{for all } x_1, x_2 \in X \ ;$$
$$\| (I-Q_o) z \|_Z \le M \| z \|_Z \quad \text{for all } z \in Z \ ;$$
$$\| (I-R') z \|_Y \le \gamma \| z \|_Z \quad \text{for all } z \in Z \cap Y_{o1} \ ;$$
$$\| Hy \|_X \le \Lambda \| y \|_Y \quad \text{for all } y \in Y_1 \subset Y \ .$$

Then, for any fixed $x^* \in X_o$, or $x^* = x_o^* + w$, $x_o^* \in X_{oo}$, $w \in \Sigma$,

$$Tx = x^* + H(I-R')(I-Q_o)Nx \in P^{-1}x^* \subset X \quad \text{for all } x \in P^{-1}x^* \ .$$

Moreover, for any two elements $x_1, x_2 \in P^{-1}x^*$, we also have

$$Tx_1 - Tx_2 = H(I-R')(I-Q_o)(Nx_1 - Nx_2) \ ,$$

$$\|Tx_1 - Tx_2\|_X \le \Lambda\gamma ML \|x_1 - x_2\|_X \ .$$

Thus, T maps $P^{-1}x^*$ into itself, and, if $k = \Lambda\gamma ML < 1$, then T is a contraction map in the norm of X.

5. <u>The use of finite elements</u>. The case of $X = C^p(\bar{G})$, $Y = C^s(\bar{G})$, $Z = C^q(\bar{G})$ for some $0 \le s < q \le p$ and a fixed bounded domain G of E^ν, $\nu \ge 1$, is of interest. Thus we assume here $N : C^p \to C^q$, $I-Q_o : C^q \to C^q$, $I-R' : C^q \to C^s$, $H : C^s \to C^p$. Now we assume that G admits of a system $\psi_1,\dots,\psi_N$ of "finite elements" of class $C^s(\bar{G})$ and "fineness" h. Let Σ' be the N-dimensional space spanned by $\psi_1,\dots,\psi_N$. Let $\phi_j = H_o\psi_j$, $j = 1,\dots,N$, be the corresponding elements in X_{o1}, and Σ the space spanned by $\phi_1,\dots,\phi_N$; thus, $\Sigma' \subset Y_{o1}$, $\Sigma \subset X_{o1}$. We conceive $R'z = \Sigma c_j\Psi_j$ as an

"approximation" of z by means of the finite system $\psi_1,\dots,\psi_N$ of fineness h. Then, for instance, for $X = C^p$, $Y = C^s$, and $z \in Z = C^q$, $0 \le s < q \le p$, we have ([25], [38]):

(10) $$\|(I - R')z\|_Y \le C\, h^{q-s}\|z\|_Z ,$$

where C is a constant. In this situation $k = \Lambda CMLh^{q-s}$, and we can make $k < 1$ by taking h sufficiently small.

(5.i) Theorem (Cesari [12]) For $X = C^p(\bar G)$, $Y = C^s(\bar G)$, $Z = C^q(\bar G)$, $0 \le s < q \le p$, the operator T defined by $Tx = Px + H(I-R')(I-Q_o)Nx$, or $T : P^{-1}x^* \to P^{-1}x^*$, $x^* = Px$ for any $x^* \in X_o$, can be made to be a contraction on the fibers $P^{-1}x^*$ of the projection operator P by taking $h > 0$ sufficiently small, and then the auxiliary equation $x = Tx$ has a unique fixed point $x = Tx = \mathcal{T}x \in P^{-1}x^*$ for any $x^* \in X_o$.

This applies to equations $Ex = Nx$ whose underlying linear problem $Ex = 0$ may be self-adjoint or nonself-adjoint with possible nontrivial null space $X_{oo} = \ker E$. Thus, even for self-adjoint problems it may not be necessary to compute eigenelements. Considerations analogous to the ones above hold also, for, e.g., $X = H^p(G)$, $Y = H^s(G)$, $Z = H^q(G)$, $0 \le s < q \le p$, Sobolev spaces.

6. A process of successive approximations. We again use the notations of §1. We assume here that the linear operator E is bounded when restricted to X_o. We also assume that there is an element $u_o \in X_o$ and positive numbers δ,η,L,J such that, if

$$\mathcal{C} = [x = u+v,\ u \in X_o,\ v \in X_1 \ \big|\ \|u-u_o\| \le \delta,\ \|v\| \le \eta] ,$$

then

(11) $\mathcal{D}(N) \supset \mathcal{C}$; $\|Nx\| \le J$ for $x \in \mathcal{C}$; and $\|Nx_1 - Nx_2\| \le L\|x_1-x_2\|$ for $x_1,x_2 \in \mathcal{C}$;

There is a bounded linear operator $D : Y_o \to X_o$ with trivial kernel such that

(12) $$K = \|I - DQE\| + \|D\|\,\|Q\|\,L < 1 ;$$

(13) $$\|H(I-Q)\| \le \ell,\ \ell J \le \eta,\ K + \ell L + \|D\|\,\|Q\|\,\ell L^2 < 1 ;$$

(14) $$\|DQ[Eu_o - N(u_o+v)]\| \le (1-K)\delta \ \text{ for all } \|v\| \le \rho .$$

A few remarks are needed here. Let $F = Q(E-N)$, and let us assume that the Fréchet derivative $\mathcal{D}Nu_o$ of N at u_o exists. Then $\mathcal{D}Fu_o = QE - Q(\mathcal{D}Nu_o)$. Let us also assume that $\mathcal{D}Fu_o$ is invertible with inverse $(\mathcal{D}Fu_o)^{-1}$. If we take $D = (\mathcal{D}Fu_o)^{-1}$ then

$$I - DQE = I - (\mathcal{D}Fu_o)^{-1}[(\mathcal{D}Fu_o) + Q(\mathcal{D}Nu_o)] = -(\mathcal{D}Fu_o)^{-1}Q(\mathcal{D}Nu_o) .$$

Because of (11) we also have $\|\mathcal{D}Nu_o\| \leq L$. Hence $\|I - DQE\| \leq \|(\mathcal{D}Fu_o)^{-1}\| \|Q\| L$, and

$$K \leq 2K_o \quad \text{with} \quad K_o = \|(\mathcal{D}Fu_o)^{-1}\| \|Q\| \|\mathcal{D}Nu_o\| .$$

Thus, instead of (12) it may be easier to verify that $K_o < 1/2$, and we may well write $(1-2K_o)\delta$ in the second member of (14).

Also we should note that

$$DQ[Eu_o - N(u_o+v)] = DQ(Eu_o-Nu_o) - DQ[N(u_o+v)-Nu_o] ,$$

$$\|DQ[Eu_o - N(u_o+v)]\| \leq \|DQ(Eu_o-Nu_o)\| + \|D\| \|Q\| L\eta .$$

Thus, instead of (14) it may be easier to verify that

$$\|DQ(Eu_o-Nu_o)\| + \|D\| \|Q\| L\eta \leq (1-K)\delta .$$

If $Q(Eu_o-Nu_o) = 0$, that is, $u_o \in X_o$ is a Galerkin approximation, then it is enough to require, instead of (14), that $\|D\| \|Q\| L\eta \leq (1-K)\delta$, or more restrictively, $\leq (1-2K_o)\delta$.

(6.i) Theorem. Under the hypotheses above, the method of successive approximations

$$x_k = u_k + v_k , \; u_k \in X_o , \; v_k \in X_1 , \; Px_k = u_k , \; v_o = 0 ,$$

$$v_k = H(I-Q)N(u_{k-1} + v_{k-1}) , \tag{15}$$

$$u_k = u_{k-1} - D[QEu_{k-1} - QN(u_{k-1} + v_k)] , \; k = 1,2,\ldots,$$

is defined, that is, $x_k \in C$ for all k , and converges toward a solution x ,that is, $u_k \to u$, $v_k \to v$, $x_k \to x = u+v$ in X as $k \to \infty$, and $Ex = Nx$. Thus, the equation $Ex = Nx$ has at least a solution in C .

This is a process of successive approximations to solutions $x \in X$ of the system (2),(3) of auxiliary and bifurcation equations in the present "nonself-adjoint" general case. It will be applied below in conjunction with finite element theory. Previous versions in more restrictive situations have been studied by Banfi [1,2], Banfi and Casadei [3], Sanchez [37]. (Cf. [21] and [31]).

Proof. First, let us prove that, if all x_k are in C and $u_k \to u$, $v_k \to v$, and $x = u+v$, then $Ex = Nx$. Indeed, the operators H,P,N,D are continous in X , E is continuous in X_o ; hence, $Px = u$,

$$v = H(I-Q)Nx , \; D[QEu - QNx] = 0 ,$$

and also $QEu - QNx = 0$. Then, by these relations and (k_{123}) we also have

$$Ev = EH(I-Q)Nx = (I-Q)Nx ,$$
$$Eu = EPu = QEu = QNx ,$$
$$Nx = QNx + (I-Q)Nx = Eu + Ev = Ex .$$

Here $(u_o,v_o) = (u_o,0) \in C$. Assume that $(u_k,v_k) \in C$ and let us prove that $(u_{k+1},v_{k+1}) \in C$. First note that $\|u_k - u_o\| \le \delta$, $\|v_k\| \le \eta$,

$$\|v_{k+1}\| = \|H(I-Q)Nx_k\| \le \|H(I-Q)\| \, \|Nx_k\| \le \ell J \le \eta ,$$

and thus $(u_k,v_{k+1}) \in C$. Now we have

$$\|v_{k+1} - v_k\| = \|H(I-Q)[N(u_k+v_k) - N(u_{k-1}+v_{k-1})]\|$$
$$\le \ell L(\|u_k - u_{k-1}\| + \|v_k - v_{k-1}\|) ,$$
$$u_{k+1} - u_k = (u_k-u_{k-1}) - DQE(u_k-u_{k-1}) + DQ[N(u_k+v_{k+1}) - N(u_{k-1}+v_k)] ,$$
$$\|u_{k+1} - u_k\| \le \|I-DQE\| \, \|u_k-u_{k-1}\| + \|D\| \, \|Q\| L[\|u_k-u_{k-1}\| + \|v_{k+1}-v_k\|]$$
$$\le K\|u_k-u_{k-1}\| + \|D\| \, \|Q\| \ell L^2(\|u_k-u_{k-1}\| + \|v_{k+1}-v_k\|)$$
$$\le (K + \|D\| \, \|Q\| \ell L^2) \|u_k-u_{k-1}\| + \|D\| \, \|Q\| \ell L^2 \|v_k-v_{k-1}\| ,$$

and finally

(16) $$\|u_{k+1}-u_k\| + \|v_{k+1}-v_k\| \le \alpha(\|u_k-u_{k-1}\| + \|v_k-v_{k-1}\|) \quad \text{for}$$
$$\alpha = K + \ell L + \|D\| \, \|Q\| \ell L^2 < 1 .$$

We also have

$$u_{k+1} - u_o = u_k - u_o - D[QEu_k-QN(u_k+v_{k+1})] + DQ[Eu_o-N(u_o+v_{k+1})] - DQ[Eu_o-N(u_o+v_{k+1})]$$
$$= (u_k-u_o) - DQE(u_k-u_o) + DQ[N(u_k+v_{k+1}) - N(u_o+v_{k+1})] - DQ[Eu_o-N(u_o+v_{k+1})] .$$

Thus,

$$\|u_{k+1} - u_o\| \le \|I-DQE\| \, \|u_k-u_o\| + \|D\| \, \|Q\| L \, \|u_k-u_o\| + \|DQ[Eu_o-N(u_o+v_{k+1})]\|$$
$$\le K \|u_k-u_o\| + (1-K)\delta \le K\delta + (1-K)\delta = \delta .$$

Thus, $(u_{k+1},v_{k+1}) \in C$. We have proved that all points (u_k,v_k) are in C , $k = 1,2,...,$ and (16) then shows that both sequences $[u_k]$, $[v_k]$ are convergent. The proof is complete.

In the situation depicted in §4 with $X_o = X_{oo} + \Sigma$, $Y_o = Y_{oo} + \Sigma'$, $E : \mathcal{D}(E) \to Y$, $(E) \subset X$, $N : X \to Z$, $I-Q_o : Z \to Z$, $I-R' : Z \to Y$, $H_o : Y_{o1} \to X_{o1}$, $H = H_o|_{Y_1}$, with elations (9) holding as stated, then ℓ in (13) can be replaced by $\Lambda\gamma M$, and then 13) becomes

$$\Lambda\gamma MJ \leq \eta \ , \ K + \Lambda\gamma ML(1 + \|D\| \ \|Q\| \ L) < 1 \ .$$

n the particular cases considered in the remarks early in this section, we may replace by the constant $2K_o$.

. <u>The use of finite elements in the process of §6.</u> As in §5 the case of $X = C^p(\bar{G})$, $= C^s(\bar{G})$, $Z = C^q(\bar{G})$ for some $0 \leq s < q \leq p$ and a fixed bounded domain G of E^ν , ≥ 1 , is of interest. Thus, we assume $N : C^p \to C^q$, $I-Q_o : C^q \to C^q$, $I-R' : C^q \to C^s$, $_o : C^s \to C^p$. Let $\psi_1,\ldots,\psi_N$ be "finite elements" in G of class $C^s(\bar{G})$, and Σ' e the N-dimensional space spanned by $\psi_1,\ldots,\psi_N$. Let $\phi_j = H_o\psi_j$, $j = 1,\ldots,N$, be he corresponding elements in X_{o1} , and Σ the space spanned by $\phi_1,\ldots,\phi_N$. Thus, $R'z = \Sigma c_j\psi_j$ is the "approximation" of z by means of the finite system $\psi_1,\ldots,\psi_N$ of fineness h , and we assume as in §5 that $\|(I-R')z\|_Y \leq Ch^{q-s}\|z\|_Z$, C a constant. In this situation, γ can be replaced by Ch^{q-s} and ℓ by ΛCMh^{q-s} . As in §6 we consider a set

$$\mathcal{C} = [x = u+v, \ u \in X_o \ , \ v \in X_1 \,|\, \|u - u_o\|_X \leq \delta, \ \|v\|_X \leq \eta] \ ,$$

where $X_o = X_{oo} + \Sigma$, $X_{oo} = \ker E$, and then the assumptions (11-14) of §6 become

(11)' $\mathcal{D}(N) \supset \mathcal{C}$; $\|Nx\|_Z \leq J$ for $x \in \mathcal{C}$, $\|Nx_1 - Nx_2\|_Z \leq L\,\|x_1-x_2\|_X$, $x_1,x_2 \in \mathcal{C}$;

there is a linear bounded operator $D : Y_o \to X_o$ with trivial kernel such that

(12)' $$K = \|I-DQE\| + \|D\| \ \|Q\| \ L < 1 \ ;$$

(13)' $$\Lambda CMJh^{q-s} \leq \eta \ , \ K + \Lambda CMLh^{q-s}(1 + \|D\| \ \|Q\| \ L) < 1 \ ;$$

(14)' $$\|DQ[Eu_o - N(u_o+v)]\|_X \leq (1-K)\delta \quad \text{for all } \|v\|_X \leq \rho \ .$$

For $F = Q(E-N)$ we may assume as in §6 that the Fréchet derivative $\mathcal{D}Nu_o$ exists and that $\mathcal{D}Fu_o = QE - Q(\mathcal{D}Nu_o)$ is invertible. Then, for $D = (\mathcal{D}Fu_o)^{-1}$ we have $K \leq 2K_o$ with $K_o = \|(\mathcal{D}Fu_o)^{-1}\| \ \|Q\| \ \|\mathcal{D}Nu_o\|$, and instead of (12)' it may be easier to verify that $K_o < 1/2$. Instead of (14)' it may be easier to verify that

$$\|DQ(Eu_o - Nu_o)\| + \|D\| \; \|Q\| \; L\,\eta \le (1-K)\delta \; .$$

If $Q(Eu_o - Nu_o) = 0$, that is, $u_o \in X_o = \ker E + \Sigma$ is a Galerkin approximation, then we may simply require, instead of (14)' that $\|D\| \; \|Q\| \; L\,\eta < (1-K)\delta$, or more restrictively $\le (1-2K_o)\delta$. Theorem (6.i) now holds under the conditions above with $X_o = X_{oo} + \Sigma$, $X_{oo} = \ker E$, $\Sigma = sp(\phi_1,\ldots,\phi_N) = H_o\Sigma'$, $\Sigma' = sp(\psi_1,\ldots,\psi_N)$, $\phi_j = H_o\psi_j$, $j = 1,\ldots,N$, where $\psi_1,\ldots,\psi_N$ is a finite system of fineness h .

Analogous statements hold of course for $X = H^p(G)$, $Y = H^s(G)$, $Z = H^q(G)$, $0 \le s < q \le p$, Sobolev spaces.

These are a few of the problems treated by Ku [31] by the method of successive approximations above and the use of finite elements:

(a) $x'' - 15x' = \arctan x + \sin 2\pi t$, $0 \le t \le 1$, $x(0) = x(1)$, $x'(0) = x'(1)$;

(b) $x'' - 15x' = x^3 - 1 + \sin 2\pi t$, $0 \le t \le 1$, $x(0) = x(1)$, $x'(0) = x'(1)$;

(c) $x_{\xi\xi} + x_{\eta\eta} = \sin x + f(\xi,\eta)$, $(\xi,\eta) \in Q = [0 \le \xi,\ \eta \le 1]$, $\partial x/\partial n = 0$ on ∂Q .

8. Quasi linear partial differential equations and systems. The Leray-Schauder well known process for the existence of solutions of quasi linear partial differential equations and systems can be expressed in terms of operational equations in a form which includes the process of §1. There are situations in which a direct method of successive approximations can be formulated and its convergence can be proved. This is the case for the boundary value problem for quasi linear hyperbolic systems considered by Cesari [9,10,11]. In these papers Cesari has given existence, uniqueness, and continuous dependence theorems for quasi linear systems of a first order partial differential equations of the Schauder type in R^{m+1} ,

$$\Sigma_{j=1}^{n} A_{ij}(x,y,z)\,[\partial z_j/\partial x + \Sigma_{k=1}^{m} c_{ik}(x,y,z)\,\partial z_j/\partial y_k] = f_i(x,y,z) \; , \; i = 1,\ldots,n \; ,$$

in a strip $D = [(x,y) \mid 0 \le x \le a,\ y = (y_1,\ldots,y_m) \in R^m]$ in the n unknowns $z(x,y) = (z_1,\ldots,z_n)$, subject to n linear boundary conditions of the form

$$\Sigma_{j=1}^{n} b_{ij}(y)\,z_j(a_i,y) = \psi_i(y) \; , \; y \in R^m \; , \; i = 1,\ldots,n \; ,$$

where $0 \le a_1,\ldots,a_n \le a$ are arbitrary numbers and $a > 0$ is sufficiently small.

This problem was motivated by questions of optical resonance (duplication of frequency) when a strong laser beam crosses a thin crystal (see [11]). A convergent method of successive approximations was formulated, and numerical results have just been obtained.

REFERENCES

1. C. Banfi, Sulla determinazione delle soluzioni periodiche di equazioni differenziali periodiche, Boll. Unione Mat. Ital. (4) 1, 1968, 608-619.

2. C. Banfi, Su un metodo di successive approssimazioni per lo studio delle soluzioni periodiche di sistemi debolmente nonlineari. Atti Accad. Sci. Torino 100, 1965-66, 471-479.

3. C. Banfi and G. Casadei, Calcolo di soluzioni periodiche di equazioni differenziali nonlineari, Calcolo 5, Suppl. 1, 1968, 1-10.

4. T.T. Bowman, Periodic solutions of Liénard systems with symmetries. To appear.

5. T.T. Bowman, The existence and bifurcation of solutions of nonlinear boundary value problems. To appear.

6. L. Cesari, Functional analysis and periodic solutions of nonlinear differential equations. Contributions to Differential Equations 1, Wiley 1963, 149-187.

7. L. Cesari, Alternative methods in nonlinear analysis. International Conference on Differential Equations (Antosiewicz ed.), Academic Press 1975, 95-148.

8. L. Cesari, Functional analysis, nonlinear differential equations, and the alternative method. (A Summer Institute at Michigan State University). Nonlinear Functional Analysis and Differential Equations (Cesari, Kannan, Schuur eds.), Marcel Dekker, New York 1976, 1-197.

9. L. Cesari, A boundary value problem for quasi linear hyperbolic systems in the Schauder canonic form. Annali Scuola Norm. Sup. Pisa (4)1, 1974, 311-358.

10. L. Cesari, A boundary value problem for quasi linear hyperbolic systems. Rivista di Matematica, Univ. di Parma (3) 3, 1974, 107-131.

11. L. Cesari, Nonlinear oscillations under hyperbolic systems. An intern. Conference, Providence, R.I.). Dynamical Systems (Cesari, Hale, LaSalle eds.), Academi Press, vol. 1, 1976, 251-261.

12. L. Cesari, Nonlinear oscillations. A Conference at the Univ. of Texas, Arlington. To appear.

13. L. Cesari, An abstract existence theorem across a point of resonance. International Symposium on Dynamical Systems, Univ. of Florida, Gainsville. Academic Press, 1977.

14. L. Cesari, Nonlinear oscillations across a point of resonance for nonself-adjoint systems. Journ. Differential Equations, to appear.

15. L. Cesari, Nonlinear problems across a point of resonance. Nonlinear Analysis, a volume in honor of E. H. Rothe, Academic Press, 1977.

16. L. Cesari and T.T. Bowman, Some error estimates by the alternative method, Quater Appl. Mathematics 35, 1977, 121-128.

17. L. Cesari and R. Kannan, Functional analysis and nonlinear differential equations Boll. Amer. Math. Soc. 79, 1973, 1216-1219.

18. L. Cesari and R. Kannan, Periodic solutions in the large of nonlinear differentia equations. Rend. Mat. Univ. Roma (2) 8, 1975, 633-654.

19. L. Cesari and R. Kannan, Solutions in the large of Liénard systems with forcing terms. Annali di Matematica pura e applicata (4) 111, 1976, 101-124.

20. L. Cesari and R. Kannan, An abstract existence theorem at resonance. Proc. Amer. Math. Soc. 63, 1977, 221-225.

21. L. Cesari and D. Ku, Alternative method and finite elements. Applicable Mathematics. To appear.

22. S.H. Chang, Existence of periodic solutions to second order nonlinear equations. Journ. Math. Anal. Appl. 52, 1975, 255-259.

23. D.C. DeFigueiredo, The Dirichlet problem for nonlinear elliptic equations: A Hilbert space approach. Partial Differential Equations and Related Topics (Dold and Eckman eds.), Springer Verlag Lecture Notes Math. 446, 1975, 144-165.

24. R. DeVries, Periodic solutions of differential systems of Lienard and Rayleigh. International Symposium on Dynamical Systems, Univ. of Florida, Gainesville. Academic Press, 1977.

25. G.J. Fix and G. Strand, An Analysis of the Finite Element Method. Prentice Hall, 1973.

26. S. Fucik, Further remarks on a theorem of Landesman and Lazer. Comm. Math. Univ. Carol. 15, 1974, 259-271.

27. J.K. Hale, Applications of alternative problems. Brown University Lecture Notes 1971, pp. 1-69.

28. J.K. Hale and R.A. Gambill, Subharmonic and ultraharmonic solutions for weakly nonlinear systems, Journ. Rat. Mech. Anal. 4, 1956, 353-398.

29. P. Hess, On a theorem by Landesman and Lazer. Indiana Univ. Math. Journ. 23, 1974, 827-829.

30. R. Kannan and P.M. McKenna, An existence theorem by alternative method for semilinear abstract equations. Boll. Unione Mat. Italiana. To appear.

31. D. Ku, Boundary value problems and numerical estimates. Univ. of Michigan thesis 1976.

. E.M. Landesman and A.C. Lazer, Nonlinear perturbations of linear boundary value problems at resonance, Journ. Math. Mech. 19, 1970, 609-623.

. A.C. Lazer and D.E. Leach, Bounded perturbations of forced harmonic oscillations at resonance. Annali di Matematica pura e applicata 72, 1969, 49-68.

. P.M. McKenna, Nonself-adjoint semilinear equations at multiple resonance in the alternative method. Journ. Differential Equations, to appear.

. K. Nagle, Boundary value problems for nonlinear ordinary differential equations, Univ. of Michigan thesis 1975.

. J. Nečas, On the range of nonlinear operators with linear asymptotic which are not invertible. Comm. Math. Univ. Carol. 14, 1973, 63-72.

. D.A. Sanchez, An iteration scheme for Hilbert space boundary value problems. Boll. Unione Mat. Italiana (4) 11, 1975, 1-9.

. R.S. Varga, Functional Analysis and Approximation Theory in Numerical Analysis. Applied Mathematics Seris, SIAM, 1971.

. S.A. Williams, A connection between the Cesari and Leray-Schauder methods. Mich. Math. Journ. 15, 1968, 441-448.

. S.A. Williams, A sharp sufficient condition for solutions of a nonlinear elliptic boundary problem. Journ. Differential Equations 8, 1970, 580-586.

ORDINARY DIFFERENTIAL SUBSPACES IN BANACH SPACES

Earl A. Coddington*
University of California, Los Angeles, U.S.A.

1. Introduction. This is an account of some recent work with A. Dijksma; the co[m]plete version will appear in [2].

Consider the eigenvalue problem

(1) $$Lf = f' = \lambda f, \quad b(f) = \int_0^1 f\, d\bar{\mu} = 0,$$

where $\mu \in BV[0,1]$. Particular cases of (1) are

(2) $$Lf = \lambda f, \quad \alpha f(0) + \beta f(1) = 0,$$

and

$$Lf = \lambda f, \quad \alpha f(0) + \beta f(1) + \gamma \int_0^1 f \bar{\varphi} = 0,$$

as well as problems involving interface conditions at a number of intermediate poi[nts] in $[0,1]$. In studying (2) in the Hilbert space $\mathfrak{H} = L^2(0,1)$ it is convenient to introduce the maximal and minimal operators, T and T_0, associated with the di[ff]erentiation operation $Lf = f'$. We identify operators in $\mathfrak{H}$ with their graphs i[n] $\mathfrak{H}^2 = \mathfrak{H} \oplus \mathfrak{H}$, so that

$$T = \{\{f,Lf\} \in \mathfrak{H}^2 \mid f \in AC[0,1]\},$$

$$T_0 = \{\{f,Lf\} \in T \mid f(0) = f(1) = 0\}.$$

Associated with (2) is the operator A given by

$$A = \{\{f,Lf\} \in T \mid \alpha f(0) + \beta f(1) = 0\},$$

and clearly $T_0 \subset A \subset T$. In analysing (2) in $\mathfrak{H}$, or equivalently A, an importa[nt] role is played by the adjoint A^*. For any linear manifold $A \subset \mathfrak{H}^2$ its adjoint [is] defined as

$$A^* = \{\{h,k\} \in \mathfrak{H}^2 \mid (g,h) - (f,k) = 0, \text{ all } \{f,g\} \in A\},$$

which is a closed linear manifold (subspace) in $\mathfrak{H}^2$. Thus, if $L^+h = -h'$, we have

$$T_0^* = \{\{h,L^+h\} \in \mathfrak{H}^2 \mid h \in AC[0,1]\},$$
$$T^* = \{\{h,L^+h\} \in T_0^* \mid h(0) = h(1) = 0\},$$
$$A^* = \{\{h,L^+h\} \in T_0^* \mid \alpha^+ f(0) + \beta^+ f(1) = 0, \alpha\overline{\alpha^+} - \beta\overline{\beta^+} = 0\},$$

and $T^* \subset A^* \subset T_0^*$.

*The work of Earl A. Coddington was supported in part by the National Science Foun[da]tion under NSF Grant MCS-76-05855.

A natural operator A associated with (1) is

$$A = \{\{f,Lf\} \in T \mid b(f) = 0\},$$

and a natural minimal operator for this problem is

$$A_0 = \{\{f,Lf\} \in T_0 \mid b(f) = 0\}.$$

Since, for $\{f,Lf\} \in T_0$ we have

$$b(f) = \int_0^1 f\, d\bar{\mu} = -\int_0^1 f'\bar{\mu} = -(f',\mu),$$

we see that

$$A_0 = \{\{f,Lf\} \in T_0 \mid (f',\mu) = 0\}.$$

Another way of writing this is

$$A_0 = T_0 \cap (B^+)^*, \qquad B^+ = \text{span}\{\{\mu,0\}\} \subset \mathfrak{H}^2.$$

Although T_0^* is (the graph of) an operator, this need not be true for A_0^*. Indeed, it can be shown that

$$A_0^* = T_0^* \dotplus B^+ = \{\{h + c\mu, -h'\} \mid \{h,-h'\} \in T_0^*,\ c \in \mathbf{C}\},$$

so that if $\mu \in AC[0,1], \mu' \in \mathfrak{H}$, we see that $\{0,\mu'\} \in A_0^*$. In general, therefore, we must think of the subspace A_0^* as a multi-valued operator, and, since $A_0 \subset A$ implies $A^* \subset A_0^*$, we must do the same for A^*.

We are thus led to consider subspaces $A \subset \mathfrak{H}^2$ and their adjoints. For n^{th} order $L = L^+$ on an arbitrary open interval $\iota = (a,b)$, we studied selfadjoint subspaces $A = A^*$ in [1], and showed how to obtain eigenfunction expansions associated with these singular problems. Here we sketch how to compute the adjoints of subspaces in general Banach spaces, and indicate applications to systems of ordinary differential operators in L^p-spaces, $1 \le p \le \infty$, and in $C(\bar{\iota})$.

2. <u>Adjoint Subspaces</u>. Let X,Y be Banach spaces over the complex field $\mathbf{C}$, and let $W = X \oplus Y$, a direct sum which is a Banach space with elements $w = \{f,g\}$, $f \in X$, $g \in Y$, and norm given via $\|w\|^2 = \|f\|^2 + \|g\|^2$. A <u>subspace</u> $A \subset W$ is a closed linear manifold considered as a linear relation (multi-valued operator), with domain

$$\mathfrak{D}(A) = \{f \in X \mid \{f,g\} \in A,\ \text{some } g \in Y\},$$

and range

$$\mathfrak{R}(A) = \{g \in Y \mid \{f,g\} \in A,\ \text{some } f \in X\}.$$

If $A(0) = \{g \in Y \mid \{0,g\} \in A\} = \{0\}$, then A is (the graph of) an <u>operator</u> from X into Y. Let X^* be the conjugate space of X, the Banach space of continuous conjugate linear functionals on X. The natural pairing between X and X^* is given by $(f,g^+) = \overline{g^+(f)}$, $f \in X, g^+ \in X$. This induces a pairing between $W = X \oplus Y$

and $W^+ = Y^* \oplus X^*$ via

$$\langle w,w^+\rangle = (g,f^+) - (f,g^+), \quad w = \{f,g\} \in W, \quad w^+ = \{f^+,g^+\} \in W^+.$$

The adjoint A^* of a linear manifold $A \subset W$ is given by

$$A^* = \{w^+ \in W^+ \mid \langle w,w^+\rangle = 0, \text{ all } w \in A\},$$

and the preadjoint $^*(A^+)$ of a linear manifold $A^+ \subset W^+$ is

$$^*(A^+) = \{w \in W \mid \langle w,w^+\rangle = 0, \text{ all } w^+ \in A^+\}.$$

It is not difficult to see that A^* is a subspace in W^+ which is weak* closed, $^*(A^+)$ is a subspace in W, and we have

$$^*(A^*) = A^c, \text{ the closure of A in W},$$

$$(^*(A^+))^* = {}^c(A^+), \text{ the weak}^* \text{ closure of } A^+ \text{ in } W^+.$$

We do not assume X,Y to be reflexive, but, in case both are, the weak* closed linear manifolds in W^+ coincide with the subspaces in W^+.

3. Adjoint pairs. If $A_0 \subset A_1$ are given subspaces in W, we consider the proble[m] of characterizing all adjoint pairs A, $A^+ = A^*$ satisfying

$$(3) \qquad A_0 \subset A \subset A_1, \quad A_0^+ = A_1^* \subset A^+ = A^* \subset A_1^+ = A_0^*.$$

This is solved in increasing detail, depending upon our knowledge of the given subspaces A_0,A_1. At the crudest level we can consider the quotient Banach spaces $V = A_1/A_0$, $V^+ = A_1^+/A_0^+$. The conjugate space of V can be identified with V^+, and we can show that $A,A^+ = A^*$ are an adjoint pair satisfying (3) if and only if A/A_0, A^+/A_0^+ are an adjoint pair.

In case $\dim(A_1/A_0) < \infty$ the adjoint pairs A,A^* satisfying (3) with given $\dim(A/A_0)$ can be described as

$$A = A_0 \dotplus M, \quad A^* = A_0^+ \dotplus M^+, \quad \text{direct sums},$$

for some appropriate M,M^+, or as

$$A = {}^*(A^*) = {}^*(A_0^+) \cap {}^*(M^+) = A_1 \cap {}^*(M^+) = \{w \in A_1 \mid \langle w,m^+\rangle = 0, \text{ all } m^+ \in M^+\},$$

$$A^* = A_0^* \cap M^* = A_1^+ \cap M^* = \{w^+ \in A_1^+ \mid \langle m,w^+\rangle = 0, \text{ all } m \in M\}.$$

The latter descriptions show how A,A^* can be viewed as being obtained from A_1,A_1^+ by imposing a finite number of generalized boundary conditions, a view which is particularly appropriate to the study of ordinary differential subspaces.

A finer characterization of adjoint pairs A, A^* with given $\dim(A/A_0)$, $\dim(A(0)/A_0(0))$, $\dim(A^*(0)/A_1^*(0))$, exists in which we can separate the description of the domains of A, A^* from that of their ranges. This simplifies considerably in case A_0, A_0^+ are operators.

Now suppose $T_0 \subset T_1 \subset W$ are given subspaces (e.g., the minimal and maximal operators for L), and let $B \subset W$, $B^+ \subset W^+$ be finite-dimensional subspaces. Then all the above results can be applied to the finite restrictions of $T_0, T_0^+ = T_1^*$ given by

$$A_0 = T_0 \cap {}^*(B^+), \quad A_0^+ = T_0^+ \cap B^*,$$

where

$$A_0^* = T_0^* \dotplus B^+, \quad {}^*(A_0^+) = T_1 \dotplus B,$$

are direct sums.

4. <u>The Case</u> $W = X \oplus X^*$. In this case $W^+ = X^{**} \oplus X^*$, and the natural imbedding of X into X^{**} implies that $W \subset W^+$. A subspace $S \subset W$ is <u>symmetric</u> if $S \subset S^*$, and $H \subset W$ is <u>selfadjoint</u> if $H = H^*$. The earlier results imply characterizations of selfadjoint extensions H of a given symmetric $S \subset W$.

A symmetric subspace $S \subset W$ is said to be <u>semibounded</u> <u>below</u> by c if $(f,g) \geq c\|f\|^2$ for all $\{f,g\} \in S$, and in this case an analog of the Friedrichs extension of a symmetric operator in a Hilbert space can be constructed.

5. <u>Ordinary differential subspaces</u>. Let $\iota = (a,b)$ be an open real interval, and let

$$L = \sum_{k=0}^{n} P_k D^k, \quad L^+ = \sum_{k=0}^{n} (-1)^k D^k P_k^*, \quad D = d/dx,$$

where the P_k are $m \times m$ matrix-valued functions of class $C^k(\iota)$, and P_n is invertible on ι. First we consider $W = X \oplus Y$, where $X = L^q(\iota)$, $Y = L^r(\iota)$, for $1 \leq q, r < \infty$. Here $L^q(\iota)$ is the Banach space of $m \times 1$ matrix-valued functions f on ι with norm

$$\|f\|_q = \left(\int_\iota |f|^q \right)^{1/q}, \quad |f|^2 = f^* f.$$

The conjugate space of X is $X^* = L^{q'}(\iota)$, where $(1/q) + (1/q') = 1$, and the pairing is given via

$$(f,g) = \int_\iota g^* f, \quad f \in L^q(\iota), \ g \in L^{q'}(\iota).$$

The Banach space $L^\infty(\iota)$ consists of all $m \times 1$ matrix-valued functions f with norm

$$\|f\|_\infty = \operatorname*{ess\,sup}_{x \in \iota} |f(x)|.$$

Minimal operators T_0, T_0^+ associated with L, L^+ can be defined as

$$T_0 = \{\{f,Lf\} \in W \mid f \in C_0^n(\iota)\}^c \subset W = L^q(\iota) \oplus L^r(\iota),$$

$$T_0^+ = {}^c\{\{f^+,L^+f^+\} \in W^+ \mid f^+ \in C_0^n(\iota)\} \subset W^+ = L^{r'}(\iota) \oplus L^{q'}(\iota),$$

and these satisfy

$$T_0 \subset T_1 = {}^*(T_0^+), \quad T_0^+ \subset T_1^+ = T_0^*,$$

and, moreover, all these subspaces are operators such that

$$\dim(T_1/T_0) = \dim(T_1^+/T_0^+) \leq 2mn,$$

with equality occurring in the regular case when a,b are finite, $P_k \in C^k(\bar{\iota})$, P_n invertible on $\bar{\iota} = [a,b]$. Thus the earlier results can be applied to finite-dimensional restrictions A_0, A_0^+ of T_0, T_0^+. In the regular case the adjoint pairs of subspaces have domains described by side conditions involving integrals as well as the usual derivatives at the end points, and such terms also enter into the description of their ranges, in addition to the differential expressions L, L^+.

The regular case is also considered in $W = X \oplus X$, where $X = C(\bar{\iota})$, the space of continuous $m \times 1$ matrix-valued functions on $\bar{\iota}$ with the sup norm. Its conjugate space is $X^* = NBV(\bar{\iota})$, the space of all $m \times 1$ matrix-valued functions μ on $\bar{\iota}$ whose components are of normalized bounded variation, with the total variation norm. The pairing is given by

$$(f,\mu) = \int_{\bar{\iota}} (d\mu^*)f, \quad f \in C(\bar{\iota}), \ \mu \in NBV(\bar{\iota}).$$

A maximal operator T_1 for L in this W is

$$T_1 = \{\{f,Lf\} \in W \mid f \in C^n(\bar{\iota})\},$$

and a minimal operator T_0 for L is obtained by restricting T_1 to those $\{f,Lf\}$ for which f and its first $n-1$ derivatives vanish at a and b. These are closed operators, but since

$$(\mathfrak{D}(T_0))^c = \{f \in C(\bar{\iota}) \mid f(a) = f(b) = 0\} \neq C(\bar{\iota}),$$

we see that T_0^* is not an operator. In fact, the central problem here is to compute T_0^*, T_1^* in $W^+ = NBV(\bar{\iota}) \oplus NBV(\bar{\iota})$. This involves a new integro-differential operator $L^\dagger$.

Finally, the regular case is considered in $W = L^q(\iota) \oplus L^r(\iota)$, where at least one of q,r is ∞. Here the maximal operator T_1 for L is

$$T_1 = \{\{f,Lf\} \in W \mid f \in C^{n-1}(\bar{\iota}), \ f^{(n-1)} \in AC(\bar{\iota})\},$$

and the minimal operator T_0 is the restriction of T_1 to those $\{f, Lf\}$ such that f and its first $n-1$ derivatives vanish at a and b. The case when $1 \leq q < \infty$, $r = \infty$ is not very different from the case when both q, r are finite. The case $q = \infty$ is more difficult, due to the fact that $(L^{\infty}(\iota))^*$ is more complicated. This case uses the methods and results from the case when $W = C(\bar{\iota}) \oplus C(\bar{\iota})$. In case $W = L^{\infty}(\iota) \oplus L^{r}(\iota)$, $1 \leq r \leq \infty$, we see that $(\mathfrak{D}(T_1))^c = C(\bar{\iota})$, which is a subspace in $L^{\infty}(\iota)$ with infinite codimension. This implies that $T_1^* = T_0^+$, $T_0^* = T_1^+$, are not operators, and, in fact, $\dim T_0^+(0) = \infty$. However, $\dim(T_1/T_0) = \dim(T_1^+/T_0^+) = 2mn$, and the abstract results apply.

References

1. E. A. Coddington and A. Dijksma, Self-adjoint subspaces and eigenfunction expansions for ordinary differential subspaces, J. Differential Equations 20 (1976), 473-526.

2. E. A. Coddington and A. Dijksma, Adjoint subspaces in Banach spaces, with applications to ordinary differential subspaces, to appear in Ann. di Mat. Pura ed Appl.

ON RELAY CONTROLLABILITY FOR BILINEAR PROCESSES

R. Conti
Istituto Matematico Ulisse Dini
Firenze, Italy

1.

Let us consider a bilinear control process (b.c.p.), i.e., a family of linear ordinary differential equations

$$\dot{x} - \left[A + \sum_{1}^{m}{}_j \, u_j \, A_j\right] x = B\,u \qquad (1.1)$$

depending on the control vector u, with components u_j .
Although b.c.p. attracted the attention of mathematicians only a few years ago, there exists already a vast and rapidly increasing literature on this subject. The book of R. R. Mohler [7] explains why b.c.p. are of importance and it gives an account on the theory and applications until 1973.
In a paper published in 1972 H. J. Sussmann [8] investigated the possibility of extending to b.c.p. the well known bang-bang and (finite bang-bang or) relay principles whose validity for linear control processes

$$\dot{x} - A\,x = B\,u$$

was established by J. P. LaSalle [6] and H. Halkin [5] ,respectively.
It is the purpose of our talk to recall Sussmann's result and to report on work done in Firenze by R. M. Bianchini and myself.

2.

First of all we have to explain symbols in (1.1).
By $u : t \mapsto u(t)$ we denote a function of $t \in \mathbb{R}$, with values $u(t) \in \mathbb{R}^m$, which is measurable and locally essentially bounded.
Every (Caratheodory) solution x of (1.1) is a function of t with values in $\mathbb{R}^n$ and, accordingly, A , A_j , B are functions of t with $n\times n$, $n\times n$, $n\times m$ matrix values respectively, which are measurable and locally integrable.

Given $\tau \in \mathbb{R}$, $\chi \in \mathbb{R}^n$, we shall denote by $x(t,\tau,\chi,u)$ the value of the solution of (1.1) corresponding to a given u and such that $x(\tau,\tau,\chi,u) = \chi$.

To a certain set U of admissible controls it corresponds, then, the attainable set at time T, i.e., the image $x(T,\tau,\chi,U)$ with respect to the mapping $u \mapsto x(T,\tau,\chi,u)$.

The general philosophy at the base of bang-bang and relay principles is that of obtaining "as much as possible" of $x(T,\tau,\chi,U)$, for a given U, by using a subset U' of controls "as simple as possible".

Let us take first as U the set

$$S = \{u \in L^\infty : |u_j(t)| \leq 1 , \quad t \in [\tau,T]\}$$

and as U' the subset S_R of relay controls, i.e., the set of $u \in S$ such that $|u_j(t)| = 1$ with a finite number of switches (= discontinuities).

A "weak" relay principle is then represented by the identity

$$\overline{x(T,\tau,\chi,S_R)} = x(T,\tau,\chi,S) \qquad (2.1)$$

($\overline{}$ denoting the closure in $\mathbb{R}^n$) proved by H. J. Sussmann [8] for bounded A, A_j and $B = 0$. The boundedness assumptions, however, can be removed by replacing the L^2-convergence arguments in Sussmann's proof by weak star convergence ones, thanks to the fact that every $u \in S$ is the weak star limit of some sequence in S_R (Cfr. G. S. Goodman [4]). Also, the assumption $B = 0$ can be removed by the same artifice which will be used in Sec. 3 (R. Conti [3]).

A "strong" relay principle, i.e.,

$$x(T,\tau,\chi,S_R) = x(T,\tau,\chi,S) \qquad (2.2)$$

can be established for a special class of b.c.p. In fact, (2.2) is valid (H. J. Sussmann [8] ; R. Conti [3]) provided that

$$A, A_j, B \text{ are piecewise analytic functions} \qquad (2.3)$$

and, further, A, A_j, B satisfy the following commutativity assumptions :

$$A(t)\,A_j(s) = A_j(s)\,A(t) \qquad (2.4)$$

$$A_j(t)\, A_k(s) = A_k(s)\, A_j(t) \qquad (2.5)$$

$$A_j(t)\, E(\tau,s)\, b^k(s) = A_k(s)\, E(\tau,t)\, b^j(t) \qquad (2.6)$$

where

$$E(t,s) = I + \int_s^t A(t_1)\, dt_1 + \int_s^t \int_s^{t_1} A(t_1)\, A(t_2)\, dt_2\, dt_1 + \dots$$

and $b^j(t)$ is the j.th column of $B(t)$.

Examples show that (2.2) may fail to be valid if anyone of the assumptions (2.3),(2.4),(2.5),(2.6) is not verified.

Recently the above results were extended by R. M. Bianchini [2] to the case of a perturbed b.c.p., i.e., to (1.1) with $B\,u$ replaced by $B\,u + c$.

3.

Let us now take as U the whole space L^∞ and as U' the set

$$R = \bigcup_{\rho \geq 0} \rho\, S_R = \{u \in L^\infty : |u_j(t)| \leq \rho\ ,\ t \in [\tau,T]\ ,\rho \geq 0,\ \text{with a finite number of switches}\}\ .$$

By the same techniques used for proving (2.2) we can obtain an "unconstrained" weak relay principle, namely

$$\overline{x(T,\tau,\chi,R)} = \overline{x(T,\tau,\chi,L^\infty)}\ . \qquad (3.1)$$

The corresponding strong principle, i.e.,

$$x(T,\tau,\chi,R) = x(T,\tau,\chi,L^\infty) \qquad (3.2)$$

can be established under the commutativity assumptions (2.4),(2.5),(2.6), omitting the analyticity assumptions (2.3) thanks to a recent result of G. Aronsson [1].

To show this, let $C : [\tau,T] \ni t \mapsto C(t)$ be a real $n \times m$ matrix valued function, measurable and integrable. We shall prove first

$$\left\{\int_\tau^T C(t)\, u(t)\, dt : u \in L^\infty\right\} = \left\{\int_\tau^T C(t)\, u(t)\, dt : u \in R\right\} \quad (3.3)$$

If we define the linear map

$$\mathcal{L} : L^\infty \ni u \mapsto \mathcal{L}u = \int_\tau^T C(t)\, u(t)\, dt \in \mathbb{R}^n$$

we have $\mathcal{L}(R) \subset \mathcal{L}(L^\infty)$, so that (3.3) is equivalent to

$$\mathcal{L}(R) \supset \mathcal{L}(L^\infty) \qquad (3.4).$$

$\mathcal{L}(L^\infty)$ is a subspace of $\mathbb{R}^n$ of dimension d , $0 \leq d \leq n$.

If $d = n$ we have $\mathcal{L}(S) \supset B_r = \{x \in \mathbb{R}^n : |x| < r\}$ for some $r > 0$.

From this it follows (G. Aronsson [1] ,Lemma 2) that $\mathcal{L}(S_R) \supset B_r$.

But then

$$\mathcal{L}(R) = \mathcal{L}(\bigcup_{\rho \geq 0} \rho S_R) = \bigcup_{\rho \geq 0} \rho \, \mathcal{L}(S_R) \supset \bigcup_{\rho \geq 0} \rho B_r = \mathbb{R}^n = \mathcal{L}(L^\infty).$$

When $1 \leq d < n$, by a suitable change of coordinates, $y = P x$,

we have

$$\left\{P^{-1}\int_\tau^T C(t)\, u(t)\, dt : u \in L^\infty\right\} = \mathbb{R}^d$$

i.e.,

$$\left\{\int_\tau^T P^{-1} C(t)\, u(t)\, dt : u \in L^\infty\right\} = \mathbb{R}^d .$$

Replacing C by $P^{-1}C$ in the first part of the proof we have

$$\left\{\int_\tau^T P^{-1} C(t)\, u(t)\, dt : u \in R\right\} \supset \left\{\int_\tau^T P^{-1} C(t)\, u(t)\, dt : u \in L^\infty\right\}$$

i.e., $P^{-1}\mathcal{L}(R) \supset P^{-1}\mathcal{L}(L^\infty)$, hence (3.4) again.

When $d = 0$, (3.3) is trivial.

From (3.3) it follows

$$\left\{\int_\tau^T \left\{\sum_1^m{}_j\, u_j(t)\, A_j(t)\right\} dt : u \in L^\infty\right\} = \left\{\int_\tau^T \left\{\sum_1^m{}_j\, u_j(t)\, A_j(t)\right\} dt : u \in R\right\} \qquad (3.5).$$

In fact we can write

$$\sum_1^m{}_j\, u_j(t)\, A_j(t) = (\, C_1(t)\, u(t) \mid \ldots \mid C_n(t)\, u(t))$$

where $C_k(t)$ is the $n \times m$ matrix whose j.th column is the k.th column of $A_j(t)$. Therefore

$$\int_\tau^T \left\{\sum_1^m{}_j\, u_j(t)\, A_j(t)\right\} dt = \left(\int_\tau^T C_1(t)\, u(t)\, dt \,\Big|\, \ldots \,\Big|\, \int_\tau^T C_n(t)\, u(t)\, dt\right)$$

and (3.3) yields (3.5).

Let $A = 0$, $B = 0$ and let (2.5) be satisfied. Then the solutions of (1.1) are represented by

$$x(T, \tau, \chi, u) = \exp\left(\int_\tau^T \left\{\sum_1^m{}_j\, u_j(t)\, A_j(t)\right\} dt\right)\chi$$

and (3.5) yields (3.2).

In the general case (1.1) can be reduced to

$$\dot{\tilde{x}}^+ - \left\{\sum_1^m{}_j\, u_j(t)\, \tilde{A}_j^+(t)\right\} \tilde{x}^+ = 0$$

where

$$\tilde{x} = E(\tau,t)\, x \ , \quad \tilde{x}^{+} = \begin{pmatrix} 1 \\ \tilde{x} \end{pmatrix}, \quad A_j^{+}(t) = \begin{pmatrix} 0 & 0 \\ E(\tau,t)\, b^j(t) & A_j(t) \end{pmatrix}$$

and (2.5) plus (2.4) and (2.6) will insure that

$$\tilde{x}^{+}(T,\tau,\chi,u) = \exp\left(\int_{\tau}^{T} \left\{ \sum_{1}^{m} {}_j\ u_j(t)\ \tilde{A}_j^{+}(t) \right\} dt \right) \tilde{\chi} \ , \quad \tilde{\chi} = \begin{pmatrix} 1 \\ \chi \end{pmatrix}.$$

Therefore $\tilde{x}^{+}(T,\tau,\chi,L^{\infty}) = \tilde{x}^{+}(T,\tau,\chi,R)$, hence $\tilde{x}(T,\tau,\chi,L^{\infty}) = \tilde{x}(T,\tau,\chi,R)$, i.e., $E(T,\tau)\, x(T,\tau,\chi,L^{\infty}) = E(T,\tau)\, x(T,\tau,\chi,R)$ and finally (3.2).

Bibliography

[1] G. Aronsson, Finite bang-bang controllability for certain non-linear systems (To appear Proc. Royal Soc. Edinburgh, Section A (Math.))

[2] R. M. Bianchini, Sul principio del bang-bang per processi di controllo bilineari affini (To appear Le Matematiche)

[3] R. Conti, Le Matematiche, 30 (1975), 365-371

[4] G. S. Goodman, From "Multiple Balayage" to Fuzzy Sets; Ist. Matem. U. Dini, Univ. di Firenze, internal report 1973/74 -40

[5] H. Halkin, J. SIAM Control, 2 (1965), 199-202

[6] J. P. LaSalle, Contributions to the Theory of Nonl. Oscillations, 5 (1960), 1-24

[7] R. R. Mohler, Bilinear Control Processes, Math. in Science and Engng., vol. 106, Academic Press, 1973

[8] H. J. Sussmann, SIAM J. Control, 10 (1972), 470-476.

SELFADJOINTNESS OF SECOND ORDER ELLIPTIC AND DEGENERATE ELLIPTIC DIFFERENTIAL OPERATORS

A. Devinatz

Northwestern University
Evanston, IL 60201/USA

1. Introduction

Consider the formally selfadjoint differential operator on R^n given by

$$(1.1) \qquad H = -\sum_{j,k=1}^{n} D_j a_{jk}(x) D_k + q(x), \quad D_k = (\partial_k - ib_k(x)),$$

where $[a_{jk}(x)]$ is a real positive definite symmetric matrix, b_k and q are real valued, and, for the moment, we assume that the a_{jk} and b_k are in $C^1(R^n)$. If in addition $q \in L^2_{loc}(R^n)$, then for every $u \in L^2$, Hu is well defined as a distribution so that we may consider $H: L^2(R^n) \to \mathcal{D}'(R^n)$, the Schwartz space of distributions. The maximal operator H_{max} associated with H is taken to be the restriction of H to those $u \in L^2$ for which $Hu \in L^2$. The minimal operator H_{min} associated with H is taken to be the closure of H restricted to $C_0^\infty(R^n)$. It is, of course, immediate that $H^*_{min} = H_{max}$.

One problem which is of physical interest, at least when $a_{jk}(x) = \delta_{jk}$, is to give conditions on q so that H_{min} is an observable; i.e., a selfadjoint operator. In [7] T. Kato, using a new differential distributional inequality was able to show, among other things, that if $a_{jk}(x) = \delta_{jk}$ and $q(x) \geq -[\alpha|x|^2 + \beta]$ $(\alpha,\beta \geq 0)$, then H_{min} is selfadjoint. This is true without any further restrictions on the functions b_k than those we noted previously.

In a subsequent paper [1], using some of the results obtained in [7], we were able to improve the result of Kato noted above, as well as a number of results which had appeared previously in the literature (see also [4] and [9]). We were also able to show that if q is bounded below, then with some growth restrictions on the a_{jk} within closed shells which expand to infinity, the operator H_{min} is selfadjoint. This shows that a theorem of Hermann Weyl for second order ordinary differential operators is "close" to being true in the higher dimensional case. Weyl's theorem is that the second order ordinary differential operator $-(pu')' + qu$, $p > 0$,

---

Research partially supported by NSF Grant MCS76-04976.

q bounded below, when restricted to $C_0^\infty(R)$ is essentially selfadjoint in $L^2(R)$. That this theorem may no longer be true for equations of the form (1.1) when $n \geq 3$ was shown by example in [10] and [12].

In the paper [8], Kato again considered the problem of selfadjointness for operators of the form (1.1) with $a_{jk} = \delta_{jk}$, but with the less restrictive condition $q \in L^1_{loc}(R^n)$. In this case there seems to be no natural way of defining a minimal operator associated with H, since H does not map $C_0^\infty(R^n)$ into $L^2(R^n)$. However, there is a natural way of defining a maximal operator associated with H, and, as Kato points out in [8], from a philosophical point of view it is as satisfactory to prove that H_{max} is selfadjoint as it is to prove that H_{min} is selfadjoint.

The first objective of this paper will be to obtain the results of [1] for the case where $q \in L^1_{loc}$. We shall do this in section 2. Next we shall consider the selfadjointness question for equations of the form (1.1) but with the important difference that the matrix $[a_{jk}(x)]$ is assumed to be only positive semi-definite. The methods used for elliptic operators are no longer available for the latter situation. In [2] we were able to get some results along this line using A. E. Nussbaum's theory of quasi-analytic vectors. In this paper we shall get somewhat more sophisticated results by the use of somewhat more sophisticated techniques, relying on the theory of stochastic differential equations. However, the results we get are still in no way comparable to the rather sharp results which are now known for elliptic operators Our results for degenerate elliptic operators will be outlined in section 3.

2. Elliptic Operators

As we noted in the introduction, the object of this section is to obtain the results of [1] for the case where $q \in L^1_{loc}$. We shall first give a generalization of Theorem 1 of [1]. Once some initial notions and facts have been noted the proof follows the one given in [1]. Further, since the initial notions and facts have been taken from Kato's paper [8], our discussion in connection with our first theorem will consist only of a brief outline.

We shall initially make the same assumptions on the coefficients of (1.1) as we made in section 1, with the exception that $q \in L^1_{loc}$ and bounded below. Following [8] we define

(2.1) $\mathscr{D}(H_{max}) = \{u \in L^2 \colon Hu \in L^2,\ qu \in L^1_{loc}\}$,

and for $u \in \mathscr{D}(H_{max})$ we define

(2.2) $$H_{max}u = Hu.$$

We also set

(2.3) $$\begin{cases} h_0[u,v] = \int \sum_{j,k} a_{jk} D_k u \overline{D_j v}, \\ h_1[u,v] = \int q u \bar{v}, \\ h = h_0 + h_1. \end{cases}$$

The domain $\mathscr{D}(h_0)$ of h_0 shall be the Cartesian product with itself of the set of elements $u \in L^2$ whose distribution derivatives are functions which satisfy $h_0[u,u] < \infty$. The domain $\mathscr{D}(h_1)$ of h_1 shall be the Cartesian product with itself of those $u \in L^2$ which satisfy $h_1[u,v] < \infty$. The domain of h shall be taken as $\mathscr{D}(h) = \mathscr{D}(h_0) \cap \mathscr{D}(h_1)$.

Using the fact that H is a strongly elliptic operator it is a straightforward exercise to show that h is a closed Hermitean symmetric form bounded below by the lower bound of q. Thus from the first fundamental theorem of quadratic forms [6; 322] there exists a selfadjoint operator T, bounded below by the lower bound of q, such that for every $u \in \mathscr{D}(T)$ and for every $v \in \mathscr{D}(h)$

(2.4) $$(Tu|v) = h[u,v].$$

Further, as is shown in [8],

(2.5) $$T \subseteq H_{max}.$$

We are now in a position to state our first theorem and indicate its proof. We make the following hypotheses on the coefficients of the operator H given by (1.1):

(i) The functions $a_{jk}(x)$ and $b_k(x)$ are real valued, the a_{jk} are locally $C^{1+\alpha}$ and the b_k are in C^1. The function $q(x)$ is real valued, bounded below and belongs to L^1_{loc}.

(ii) The matrix $[a_{jk}(x)]$ is symmetric and positive definite for each $x \in R^n$.

(iii) There exist fixed positive numbers e, r, and K, and a sequence $\{\Omega_k\}$ of bounded domains in R^n with the property that for any bounded set $B \subset R^n$ there is an integer k so that $B \subset \Omega_k$, and such that the following is true: For each k and for each $y \in \partial\Omega_k$ there is a positive function $a_y(x)$ of class $C^{2+\alpha}$ defined on the ball $B_y = \{x \colon |x - y| < r\}$ so that

1° all of the functions $a_{jk}(x)a_y(x)$, $a_{jk}(x)\partial_j a_y(x)$ and their first derivatives have K as a common bound and a common Hölder constant, and

2° for every $x \in B_y$ and for every $\xi \in R^n$

$$a_y(x) \sum_{j,k=1}^{n} a_{jk}(x)\xi_j\xi_k \geq e|\xi|^2.$$

Theorem 1. Under the hypotheses (i) to (iii) the operator H_{max} is selfadjoint.

Proof. It follows from (2.5) that if it can be shown that $\mathcal{D}(H_{max}) \subseteq \mathcal{D}(T)$, then the theorem is proved. We first reproduce an argument due to Kato [8]. Suppose $u \in \mathcal{D}(H_{max})$ and set $v = (T+\alpha)^{-1}(H_{max}+\alpha)u$, where α is a real number chosen sufficiently large so that $(T+\alpha)^{-1}$ exists on L^2. Thus $(T+\alpha)v = (H_{max}+\alpha)u$ and hence by (2.5), $(H+\alpha)w = (H_{max}+\alpha)w = 0$, where $w = u - v$. Thus $\tilde{H}w = qw \in L^1_{loc}$ since $w \in \mathcal{D}(H_{max})$, where $\tilde{H} = -H + q$. If H_0 is the operator $\tilde{H}$, with each b_k set equal to zero, then it follows from Kato's inequality [7], that $H_0|w| \geq q|w|$. If we suppose that $q \geq 0$, as we may do without loss of generality, we get $H_0|w| \geq 0$. The proof of Theorem 1 of [1], using Littman's general maximum principle, shows that $w = 0$. Thus $u = v \in \mathcal{D}(T)$ and the proof is complete.

Our next objective is to obtain a more general version of Theorem 2 of [1]. The proof is basically the same as that given in [1]. However, since the proof requires a slightly different approach, and is in some sense technically simpler, we shall provide some of the details. In what follows we shall take

$$\lambda(r) = \sup\{\sum_{j,k} a_{jk}(x)\xi_j\xi_k : |\xi| = 1,\ |x| = r\}. \tag{2.6}$$

Theorem 2. Assume the hypotheses (i) and (ii) of Theorem 1 with the exception that q is locally bounded below. Assume further that there exists a non-negative function $\chi \in C^1([0,\infty))$ such that

$$\int_0^\infty \chi\lambda^{-1/2} = \infty, \tag{2.7}$$

$$\chi(t) \leq a[\int_0^t \chi\lambda^{-1/2}] + b, \tag{2.8}$$

$$\chi^2(|x|)q(x) \geq -c[\int_0^{|x|} \chi\lambda^{-1/2}]^2 - d + \gamma \sum_{j,k} a_{jk}(x)\frac{x_j x_k}{|x|^2}[\chi'(|x|)]^2, \tag{2.9}$$

where $1 < \gamma < \infty$, and a,b,c and d are positive constants. Then H_{max} is a closed symmetric operator.

Before we proceed with the proof we shall make some comments and establish some auxiliary facts. First let us notice that the conditions (2.7) and (2.8) imply that

(2.10) $$\int_0^\infty \lambda^{-1/2} = \infty.$$

We shall not use this fact in the sequel, but it does show that the hypotheses of Theorem 2 impose considerably more severe growth conditions on the principal coefficients than imposed by Theorem 1.

To prove (2.10), assume to the contrary that this integral is finite. Let $t_n \in [0,n]$ so that

$$\chi(t_n) = \sup\{\chi(t): t \in [0,n]\}.$$

If χ is unbounded, $\chi(t_n) \uparrow \infty$ as $n \to \infty$. Thus from (2.8) we have

(2.11) $$1 \leq \frac{a}{\chi(t_n)} \int_0^{t_n} \chi\lambda^{-1/2} + \frac{b}{\chi(t_n)} .$$

Since $\chi(t) \leq \chi(t_n)$ for $t \in [0,n]$, and since $\lambda^{-1/2} \in L^1$, it follows that

$$\frac{1}{\chi(t_n)} \int_0^{t_n} \chi\lambda^{-1/2} \to 0 \quad \text{as } n \to \infty.$$

But this contradicts (2.11). Thus χ is bounded. But then the integral in (2.7) converges, which is again a contradiction.

Before we proceed to the proof of Theorem 2 let us first establish a lemma.

<u>Lemma 2.1</u>. <u>There exists a positive constant</u> C <u>so that for every</u> $r \geq 0$ <u>there exists an</u> $s > r$ <u>and</u> $\phi \in C^2(R)$ <u>so that</u> $\phi(t) = 1$ <u>for</u> $t \leq r$, $0 \leq \phi \leq 1$, $\phi(t) = 0$ <u>for</u> $t \geq s$, <u>and</u>

$$|\phi'(t)| \leq \frac{C\chi(t)\lambda^{-1/2}(t)}{\int_0^s \chi\lambda^{-1/2}}, \quad t \geq 0.$$

Proof. For the purposes of this proof we may suppose that λ is differentiable, since otherwise we could work with a differentiable function larger than λ, but satisfying the condition (2.7). Let us set $w(s) = \int_0^s \chi\lambda^{-1/2}$. Since $w(s) \to \infty$ as $s \to \infty$ it is clear that given r there exists an $s > r$ so that

$$\int_r^s \chi\lambda^{-1/2} = \frac{1}{2} w(s).$$

Let $0 < \eta \leq w(s)/4$ and let $\mu > 0$ so that

$$\int_{r+\mu}^{s-\mu} \chi\lambda^{-1/2} \geq \frac{1}{2} w(s) - \eta.$$

Let $\zeta \in C_0^1([r,s])$ so that $0 \leq \zeta \leq 1$, and $\zeta \equiv 1$ on $[r+\mu, s-\mu]$. Then

$$\frac{1}{2}\,w(s) - \eta \le \int_r^s \zeta\chi\lambda^{-1/2}.$$

Let us set

$$\Phi = \frac{\zeta\chi\lambda^{-1/2}}{\int_r^s \zeta\chi\lambda^{-1/2}} \le \frac{\chi\lambda^{-1/2}}{\frac{1}{2}\,w(s) - \eta} \le \frac{4\chi\lambda^{-1/2}}{w(s)}.$$

Clearly, $\int_r^s \Phi = 1$.

Now, let us set

$$\phi(t) = \begin{cases} 1 & \underline{\text{for}}\ 0 \le t \le r, \\ 1 - \int_r^t \Phi & \underline{\text{for}}\ r \le t \le s, \\ 0 & \underline{\text{for}}\ t \ge s. \end{cases}$$

The function ϕ satisfies the conditions of the corollary.

Proof of Theorem 2. Let B be a ball in R^n with center at the origin and let H_B be the maximal operator corresponding to the differential operator obtained from H by changing, if necessary, the coefficients a_{jk}, b_j and q outside of B so that the conditions of Theorem 1 are fulfilled. Thus H_B is a selfadjoint operator. If $\phi \in C_0^2(B)$ and $u \in \mathcal{D}(H_{max})$ then as shown in [7], $\phi u \in \mathcal{D}(H_B)$. Thus it follows from (2.4) with $T = H_B$ and $v = \chi^2 \phi u$ that

$$\int \phi\chi^2 \bar{u} H_{max} \phi u = \int a_{jk} D_k \phi u\, \overline{D_j \chi^2 \phi u} + \int \chi^2 \phi^2 q |u|^2, \tag{2.12}$$

where we are using the summation convention on the right. From (2.12) it follows by straightforward computations that

$$\int \phi^2\chi^2 a_{jk} D_k u \overline{D_j u} + \int \phi^2\chi^2\, q|u|^2 = \int \phi^2\chi^2\, \bar{u} H_{max} u + \sum_{k=1}^{5} I_k, \tag{2.13}$$

where

$$I_1 = -2(1+\mathrm{Re})\int \chi^2 \phi \bar{u} a_{jk} \partial_j \phi D_k u, \qquad I_4 = -\int \chi^2 \phi |u|^2 \partial_j [a_{jk} \partial_k \phi],$$

$$I_2 = -2\int \phi^2 \chi\, \chi' \bar{u} a_{jk} \frac{x_j}{|x|} D_k u, \qquad I_5 = -\int \chi^2\, |u|^2 a_{jk} \partial_j \phi \partial_k \phi.$$

$$I_3 = -2\int \phi |u|^2\, \chi\chi' a_{jk} \frac{x_j}{|x|}\, \partial_k \phi,$$

We wish to obtain estimates for the integrals I_k. To estimate I_1 we first apply the Schwarz inequality for the quadratic form $a_{jk}\xi_j\xi_k$ and then apply the Schwarz inequality to the integral to get

$$|I_1| \le 4\{\int \phi^2\chi^2 a_{jk} D_k u \overline{D_j u}\}^{1/2} \{\int \chi^2 |u|^2 a_{jk} \partial_k \phi \partial_j \phi\}^{1/2}$$

$$\le \varepsilon_1 \int \phi^2\chi^2 a_{jk} D_k u \overline{D_j u} + \frac{4}{\varepsilon_1} \int \chi^2 |u|^2 a_{jk} \partial_k \phi \partial_j \phi, \qquad \varepsilon_1 > 0.$$

The integrals I_2 and I_3 can be handled in the same way and we get the estimates

$$|I_2| \le \varepsilon_2 \int \phi^2\chi^2 a_{jk} D_k u \overline{D_j u} + \frac{1}{\varepsilon_2} \int \phi^2 (\chi')^2 |u|^2 a_{jk} \frac{x_j x_k}{|x|^2}, \quad \varepsilon_2 > 0,$$

$$|I_3| \le \varepsilon_3 \int \phi^2 |u|^2 (\chi')^2 a_{jk} \frac{x_j x_k}{|x|^2} + \frac{1}{\varepsilon_3} \int \chi^2 |u|^2 a_{jk} \partial_k \phi \partial_j \phi, \quad \varepsilon_3 > 0.$$

Finally we estimate the sum $I_4 + I_5$. Taking into account the fact that $u \in H^1_{loc}$ we may integrate by parts to get

$$-\int \chi^2 \phi |u|^2 \partial_j [a_{jk} \partial_k \phi] = \sum_{k=1}^{3} I_{4k},$$

where

$$I_{41} = \int \chi^2 \phi a_{jk} \partial_j |u|^2 \partial_k \phi,$$

$$I_{42} = \int \phi \chi \chi' |u|^2 a_{jk} \frac{x_j}{|x|} \partial_k \phi,$$

$$I_{43} = \int \chi^2 |u|^2 a_{jk} \partial_j \phi \partial_k \phi.$$

The integral I_{43} cancels I_5, and we can handle I_{42} as previously to get

$$|I_{42}| \le \varepsilon_4 \int (\chi')^2 \phi^2 |u|^2 a_{jk} \frac{x_j x_k}{|x|^2} + \frac{1}{\varepsilon_4} \int \chi^2 |u|^2 a_{jk} \partial_j \phi \partial_k \phi, \quad \varepsilon_4 > 0.$$

To estimate the integral I_{41} we first note that

$$\partial_j |u|^2 = u \partial_j \bar{u} + \bar{u} \partial_j u = u \overline{D_j u} + \bar{u} D_j u.$$

If we use this and proceed as before we get

$$|I_{41}| \le \varepsilon_5 \int \phi^2 \chi^2 a_{jk} D_k u \overline{D_j u} + \frac{1}{\varepsilon_5} \int \chi^2 |u|^2 a_{jk} \partial_j \phi \partial_k \phi, \quad \varepsilon_5 > 0.$$

Thus we have an estimate for $I_4 + I_5$.

If we use the estimates we have obtained above in (2.13) we get

$$(1-\eta) \int \phi^2 \chi^2 a_{jk} D_k u \overline{D_j u} \le \int \phi^2 \{-\chi^2 q + \gamma (\chi')^2 a_{jk} \frac{x_j x_k}{|x|^2}\} |u|^2 \tag{2.14}$$
$$+ \int \phi^2 \chi^2 |\bar{u} H_{max} u| + C \int \chi^2 |u|^2 a_{jk} \partial_j \phi \partial_k \phi,$$

where $\eta = \varepsilon_1 + \varepsilon_2 + \varepsilon_5$, $\gamma = (1/\varepsilon_2) + \varepsilon_3 + \varepsilon_4$, and $C = (4/\varepsilon_1) + (1/\varepsilon_3) + (1/\varepsilon_4) + (1/\varepsilon_5)$. If $0 < \varepsilon_2 < 1$, we may choose ε_1 and ε_5 small enough so that $1-\eta > 0$. Further for any γ, $1 < \gamma < \infty$, we may always find numbers $0 < \varepsilon_2 < 1$, $\varepsilon_3 > 0$ and $\varepsilon_4 > 0$ so that $\gamma = (1/\varepsilon_2) + \varepsilon_3 + \varepsilon_4$.

If we use the conditions (2.8) and (2.9) in the inequality (2.14) we find that there is a positive constant C so that

$$\int\phi^2\chi^2 a_{jk}D_k u\overline{D_j u} \le C\int\phi^2[a^2w^2+b^2]|uH_{max}u| + C\int\phi^2[cw^2+d]|u|^2$$

$$+ C\int[a^2w^2+b^2]|u|^2 a_{jk}\partial_j\phi\partial_k\phi, \tag{2.15}$$

where $w(|x|) = \int_0^{|x|}\chi^2\lambda^{-1/2}$.

We shall now choose a ϕ as constructed in Lemma 2.1 and consider $\phi(|x|)$ as entering in the inequality (2.15). Thus, given r there is an $s > r$ so that if ϕ_{rs} is the function of Lemma 2.1 we have

$$a_{jk}(x)\partial_j\phi_{rs}(|x|)\partial_k\phi_{rs}(|x|) \le \frac{C\chi^2(|x|)\lambda^{-1}(|x|)}{w(s)} a_{jk}(x)\frac{x_j x_k}{|x|^2},$$

where $w(s) = \int_0^s \chi^2\lambda^{-1/2}$. Given $u \in \mathcal{D}(H_{max})$ and $\varepsilon > 0$ we see from (2.15) that since $|u|^2$ and $uH_{max}u$ are in L^1, if r is sufficiently large, then

$$\int\phi_{rs}^2\chi^2 a_{jk}D_k u\overline{D_j u} \le \varepsilon w^2(s). \tag{2.16}$$

It is now an easy matter to finish the proof of symmetry of H_{max}. We use the formula (2.12) with $\chi \equiv 1$. Let $\{r_m\}$ be a sequence increasing to ∞, $\{s_m\}$ the corresponding sequence of Lemma 2.1 and $\phi_m = \phi_{r_m s_m}$ the functions constructed in that lemma. The equality (2.12) becomes

$$\int a_{jk}D_k\phi_m u\overline{D_j\phi_m u} + \int\phi_m^2 q|u|^2$$

$$= \int\phi_m^2\bar{u}H_{max}u - 2\int\phi_m\bar{u}a_{jk}\partial_j\phi_m D_k u - \int\phi_m|u|^2\partial_j[a_{jk}\partial_k\phi_m]. \tag{2.17}$$

Using (2.16) and the estimate on the derivative of ϕ_m we get

$$|2\int\phi_m\bar{u}a_{jk}\partial_j\phi_m D_k u| \le C\int\phi_m|u|\frac{\chi\lambda^{-1/2}}{w(s_m)}|a_{jk}\frac{x_j}{|x|}D_k u|$$

$$\le \{\int\frac{\phi_m^2\chi^2}{w^2(s_m)}a_{jk}D_k u\overline{D_j u}\}^{1/2}\{\int|u|^2\lambda^{-1}a_{jk}\frac{x_j x_k}{|x|^2}\}^{1/2} = o(1) \quad \text{as } m \to \infty.$$

The last integral on the right hand side of (2.17) may be handled as before by integrating by parts. We get

$$-\int\phi_m|u|^2\partial_j[a_{jk}\partial_k\phi_m] = \int\phi_m a_{jk}[u\overline{D_j u}+\bar{u}D_j u]\partial_k\phi_m + \int|u|^2 a_{jk}\partial_j\phi_m\partial_k\phi_m.$$

Proceeding as above we see that the first integral on the right is $o(1)$ as $m \to \infty$. An easy estimation shows that this is also true for the second integral on the right. Thus from (2.17) we see that

$$\lim_{m\to\infty}[\int a_{jk}D_k\phi_m u\overline{D_j\phi_m u} + \int \phi_m^2 q|u|^2] = (H_{max}\, u|u).$$

Since the left hand side is real we see that H_{max} is symmetric.

It remains to prove the closure of H_{max}. The proof is adapted from Kato [8]. We shall give a few details. Let B be a ball in R^n and set, for $u \in H^1_{loc}$,

$$\| u \|_{1,B} = \int_B [\sum_k |D_k u|^2 + |u|^2].$$

If q is locally bounded below then $\mathcal{D}(H_{max}) \subseteq H^1_{loc}$ and $u \in \mathcal{D}(H_{max})$ implies

(2.18) $$\| u \|_{1,B} \leq C_B(\| H_{max}\, u\| + \| u\|).$$

As we have noted before $u \in C_0^2(2B)$ implies $\phi u \in \mathcal{D}(H_{2B})$ so that $u \in H^1_{loc}$. The inequality (2.18) is an immediate consequence of (2.14) with $\chi = 1$ and $\phi = 1$ in B.

Let $\{u_n\} \subseteq \mathcal{D}(H_{max})$ so that $u_n \to u$, $H_{max}\, u_n \to v$ in L^2. We must show that $u \in \mathcal{D}(H_{max})$ and $H_{max} u = v$. From (2.18) it follows that $\{u_n\}$ is Cauchy in H^1_{loc} and hence must converge to an element of that space. Clearly this element must coincide with u so that $u \in H^1_{loc}$. Now,

$$H_{2B}(\phi u_n) = \phi H_{max}\, u_n - 2a_{jk}\partial_j\phi D_k u_n - u_n\partial_j[a_{jk}\partial_k\phi].$$

We see from this that $H_{2B}(\phi u_n)$ converges in L^2 as $n \to \infty$. Since H_{2B} is closed, $\phi u \in \mathcal{D}(H_{2B})$ and

$$H_{2B}(\phi u) = \phi v - 2a_{jk}\partial_j\phi D_k u - u\partial_j[a_{jk}\partial_k\phi].$$

If we take $\phi = 1$ on B, then for every $x \in B$ we get

$$H_{max}\, u(x) = v(x).$$

Thus $u \in \mathcal{D}(H_{max})$ and H_{max} is closed. The proof of the theorem is complete.

We have not, in general, been able to prove that the differential operator of Theorem 2 is selfadjoint. However, by following the proof in Kato [8] we can show that it is selfadjoint under some additional restrictions on the coefficients a_{jk}.

Corollary 2.1. *Under the hypotheses of Theorem 2 (with $\chi = 1$) and under the additional hypothesis*

(2.19) $$\sum_{j,k}\{|a_{jk}(x)| + |\partial_j a_{jk}(x)|^2\} \leq K(1 + |x|^2),$$

the operator H_{max} is selfadjoint.

Proof. Let $\{r_m\}$ be a sequence of real numbers going to ∞. Let B_m be the ball about the origin of radius r_m^2 and let H_{B_m} be the selfadjoint maximal differential operator obtained by changing the coefficients of H outside of B_m in an appropriate way. Let $f \in L^2$ and $u_m \in \mathcal{D}(H_{B_m})$ so that

$$(H_{B_m} + i)u_m = f.$$

It is clear that

$$(2.20) \qquad \| u_m \| \leq \| f \|, \qquad \| H_{B_m} u_m \| \leq 2\| f \|.$$

Let $\phi_m \in C_0^2(B_m)$ so that $\phi_m = 1$ for $|x| \leq r_m$, $\phi_m = 0$ for $|x| \geq r_m^2$. Since $H\phi_m^2 u_m = H_{B_m}\phi_m^2 u_m \in L^2$, it follows that $\phi_m^2 u_m \in \mathcal{D}(H_{max})$. Further

$$(H_{max} + i)\phi_m^2 u_m = (H_{B_m} + i)\phi_m^2 u_m = \phi_m^2 f - 2a_{jk}\partial_j \phi_m^2 D_k u_m - u_m \partial_j [a_{jk}\partial_k \phi_m^2].$$

Let $v \in L^2$ and vanish outside of a compact set. Then for all sufficiently large m

$$((H_{max} + i)\phi_m^2 u_m | v) = (f|v).$$

From this it follows that if $\{a_{jk}\partial_j \phi_m^2 D_k u_m\}$ and $\{u_m \partial_j [a_{jk}\partial_k \phi_m^2]\}$ are sets of uniformly bounded linear functionals on L^2, then $(H_{max} + i)\phi_m^2 u_m \to f$ weakly in L^2. But H_{max} is a closed symmetric operator so that the range of $(H_{max} + i)$ is strongly closed and hence weakly closed. Thus the range of $(H_{max} + i)$ is all of L^2, which makes H_{max} selfadjoint.

In order to establish the facts about the sets of functions $\{a_{jk}\partial_j \phi_m^2 D_k u_m\}$ and $\{u_m \partial_j a_{jk}\partial_k \phi_m^2\}$ we take the functions ϕ_m in a special way. Let $r > 4$ and let $\zeta \in C_0^1(R)$ so that $\zeta(t) = 1$ for $2r < t < r^2/2$, $0 \leq \zeta(t) \leq 1$, $|\zeta'(t)| \leq C/r$ for $r \leq t \leq 2r$, $|\zeta'(t)| \leq C/r^2$ for $r^2/2 \leq t \leq r^2$ and $\zeta(t) = 0$ outside of $[r, r^2]$. Define

$$\Phi(t) = \zeta(t)/t \Big/ \int_r^{r^2} \zeta(t)/t.$$

Then $\Phi \in C_0^1(R)$ and $\int_r^{r^2} \Phi = 1$. Now define

$$\phi(t) = \begin{cases} 1 & \underline{\text{for}}\ t \leq r, \\ 1 - \int_r^t \Phi & \underline{\text{for}}\ r \leq t \leq r^2, \\ 0 & \underline{\text{for}}\ t \geq r^2. \end{cases}$$

Clearly $0 \leq \phi(t) \leq 1$ and easy estimations show that

$$(2.21) \qquad \phi'(t) = O\left(\frac{1}{t\ \ln r}\right), \qquad \phi''(t) = O\left(\frac{1}{t^2 \ln r}\right).$$

We shall take $\phi_m(t)$ to be the function constructed above when $r = r_m$, and (with a slight abuse of notation) $\phi_m(x) = \phi_m(|x|)$.

To show that the sequence $\{a_{jk}D_k u_m D_j \phi_m^2\}$ is uniformly bounded in L^2 we have

(2.22) $$\int |a_{jk}D_k u_m \partial_j \phi_m^2|^2 \le 4\int \phi_m^2 [a_{jk}\partial_j \phi_m \partial_k \phi_m][a_{jk}D_k u_m \overline{D_j u_m}].$$

Using the hypothesis (2.19) and the first estimate of (2.22) it is immediate that $(\ell n\ r_m)^2 a_{jk}\partial_j \phi_m \partial_k \phi_m$ is bounded, uniformly in m. From (2.14), with $\chi = 1$, we have

(2.23) $$\int \phi_m^2 a_{jk}D_k u_m \overline{D_j u_m} \le \int \phi_m^2 |\bar{u}_m H_{max}\, u_m| + C\int |u_m|^2 a_{jk}\partial_j \phi_m \partial_k \phi_m - \int \phi_m^2 q|u_m|^2.$$

But from the hypothesis (2.9), with $\chi = 1$, and the hypothesis (2.19) we have

$$-q(x) \le a\Big[\int_1^{|x|} \frac{dt}{t}\Big]^2 + b = a\ \ell n|x|^2 + b \le C(1 + (\ell n\ r_m)^2) \quad \text{for } |x| \le r_m^2.$$

Using this in (2.23) and recalling (2.20) we get

$$\int \phi_m^2 a_{jk}D_k u_m \overline{D_j u_m} \le C|\| u_m \| \| H_{max}\, u_m \| + (\ell n\ r_m)^2 \| u_m \|^2| \le C(\ell n\ r_m)^2 \| f \|.$$

If we use this in (2.22) we have the uniform boundedness of the left hand side.

Finally, we note that

$$\partial_j [a_{jk}\partial_k \phi_m^2] = 2\phi_m [a_{jk}\partial_{jk}^2 \phi_m + (\partial_j a_{jk})\partial_k \phi_m].$$

From the hypothesis (2.19) and the estimates (2.21) these functions are not only uniformly bounded but go to zero uniformly as $m \to 0$. Since $\| u_m \| \le \| f \|$, $u_m \partial_j [a_{jk}\partial_k \phi_m] \to 0$ in L^2 as $m \to \infty$. The proof of the corollary is complete.

Remark. Note that the proof of Corollary 2.1 remains unchanged if (2.19) is satisfied only on a sequence of intervals $\{[r_m, r_m^2]\}$, where $r_m \to \infty$.

3. Degenerate-Elliptic Operators

When the matrix $[a_{jk}(x)]$ in (1.1) is only positive semi-definite, then the methods of section 2 are no longer available. We shall use a different technique employing the theory of semi-groups and the theory of stochastic differential equations. The results obtained in this way are not nearly as sharp as for the elliptic case. The method we use is based, in part, on a simple and elegant lemma, due essentially to E. Nelson [11]. We shall state and prove a version of his lemma which is needed for our considerations.

Lemma 3.1. Let H be a closed, symmetric, negative-definite operator in a Hilber space $\mathfrak{H}$. Let $\{T_t\}$ be a C_0 semi-group on $\mathfrak{H}$, and $\mathcal{D}$ a dense set in $\mathcal{D}(H)$ so that $u \in \mathcal{D}$ implies $T_t u \in \mathcal{D}(H)$ and $\partial T_t u/\partial t = HT_t u$. Then H is selfadjoint.

Proof. Since $H \leq 0$, it follows that for any $\alpha > 0$, if the only solution in $\mathfrak{H}$ to $(H^* - \alpha I)v = 0$ is $v = 0$, then the deficiency index of H is (0,0) and H is selfadjoint. Let $u \in \mathcal{D}$, $\alpha > 0$, and $v \in \mathfrak{H}$ so that $(H^* - \alpha I)v = 0$. Let us set

$$f(t) = (e^{-\alpha t}\, T_t u/v). \tag{3.1}$$

Differentiating with respect to t gives

$$f'(t) = (e^{-\alpha t}(H - \alpha I)I_t u/v) = (e^{-\alpha t}\, T_t u/(H^* - \alpha I)v) = 0.$$

Thus f is a constant function. Now, there exist constants C and ω so that $\|T_t\| \leq C \exp \omega t$. Thus if $\alpha > \omega$ it follows from (3.1) that $f(t) \to 0$ as $t \to \infty$, so that $f(t) = 0$. But letting $t \to 0$ in (3.1) shows that $(u/v) = 0$. Since $\mathcal{D}$ is dense in $\mathfrak{H}$, $v = 0$.

We shall first consider a potential free operator of the form

$$H = \sum_{j,k} \partial_j a_{jk}(x)\partial_k. \tag{3.2}$$

In order to indicate the simple idea behind the proof of selfadjointness of the minimal operator associated with H, we shall make unnecessarily stringent hypotheses on the matrix $[a_{jk}(x)]$. We shall make the following assumptions:

$$\sum_{j,k} a_{jk}(x)\xi_j\xi_k \geq 0, \quad \xi \in R^n. \tag{3.3}$$

$$[a_{jk}(x)] = \tfrac{1}{2}\,\sigma(x)\sigma^*(x), \quad \text{where } \sigma \in C^3(R^n) \tag{3.4}$$

and

$$|\nabla\sigma(x)| \leq K, \quad |\nabla^3\sigma(x)| \leq K(1 + |x|^m). \tag{3.5}$$

Let us now consider the stochastic initial value problem

$$\begin{aligned} d\xi(t) &= b(\xi(t))dt + \sigma(\xi(t))dw(t), \quad t \geq 0, \\ \xi(0) &= x, \end{aligned} \tag{3.6}$$

where $b = (b_1,\dots,b_n)$, $b_k = \sum_j \partial_j a_{jk}(x)$, and $w(t)$ is an n-dimensional Wiener process. The assumptions (3.5) imply that

$$|b(x)| + |\sigma(x)| \leq K(1 + |x|), \quad |\nabla^2 b(x)| + |\nabla^2\sigma(x)| \leq K(1 + |x|^m). \tag{3.7}$$

Further, the differentiability condition on σ implies that b and σ are locally Lipschitz. Under these conditions it is well known [5] that the problem (3.6) has a unique solution (a.s.) which we shall call $\xi_x(t)$.

Let u be a bounded measurable function on R^n. The semi-group associated with the stationary Markov process $\xi_x(t)$ is given by

$$T_t u(x) = E[u(\xi_x(t))],$$

where, of course, E is the expectation. $\{T_t\}$ is a contraction C_0 semi-group acting on L^∞ to L^∞. As is well known [5], the differential generator of $\{T_t\}$ is the operator H of (3.2). That is to say, if $u \in C_0^2(R^n)$, then u is in the domain of the infinitesimal generator of T_t and

$$\left.\frac{\partial T_t u(x)}{\partial t}\right|_{t=0} = Hu(x).$$

Indeed, even more than this, the conditions (3.7) imply [5] that for $u \in C_0^2$, $T_t u(x)$ is continuously differentiable in t, twice continuously differentiable in x and

$$\frac{\partial T_t u(x)}{\partial t} = HT_t u(x). \tag{3.8}$$

Using (3.8) it is not hard to show by the "energy inequality method" that $\{T_t\}$ is a bounded semi-group on L^2. Indeed, set $u_t = T_t u$ and write out the equality (3.8) for $u_t \exp -ct$ instead of u_t. Multiply the resulting equation by $\phi u_t \exp -ct$, where $\phi \in C_0^2$, and then integrate over $[0,t] \times R^n$. A few integration by parts on the right with respect to the x variable ultimately lead to the inequality

$$\int_{R^n} \phi\{e^{-2ct}|u_t|^2 - |u|^2\}dx \leq \int_0^t e^{-2cs} \int_{R^n} |u_s|^2\{\partial_j[a_{jk}\partial_k\phi] - c\}dxds. \tag{3.9}$$

Now let $r_m \to \infty$, and let ϕ_m be the corresponding functions constructed in Corollary 2.1. The condition (3.5) implies that

$$|\partial_j a_{jk}(x)|^2 + |a_{jk}(x)| \leq K(1 + |x|^2).$$

Hence if we use $\phi_m(|x|)$ in place of ϕ in (3.9), as we have noted in section 2, $\partial_j[a_{jk}\partial_k\phi_m] \to 0$ uniformly as $m \to \infty$. This means that for every $c > 0$ the right hand side of (3.9) is non-positive for all sufficiently large m. Thus $\|T_t\| \leq 1$. The fact that $\{T_t\}$ is a C_0 semi-group follows easily from the considerations in [3;353].

In order to apply Nelson's Lemma 3.1, with $\mathcal{D} = C_0^2(R^n)$, and $\{T_t\}$ the semi-group given above we must show that $T_t C_0^2 \subseteq \mathcal{D}(H_{min})$. To do this let us take ϕ_m as above, and let us set $u_t = T_t u$. Now, $\phi_m^2 u_t \in \mathcal{D}(H_{min})$. This is easily shown by first mollifying u_t to a C^∞ function. A straightforward computation gives

$$(3.10) \qquad H\phi_m^2 u_t = \phi_m^2 H u_t + 2a_{jk}\partial_j u_t \partial_k \phi_m^2 + u_t \partial_j [a_{jk}\partial_k \phi_m^2].$$

The same computations as made in the proof of Corollary 2.1 show that $a_{jk}\partial_j u_t \partial_k \phi_m \to 0$ in L^2 as $m \to \infty$. In Corollary 2.1 we were only able to get weak convergence in L^2 because of the presence of a potential with a possibly unbounded negative part. Also, in the same way as in the proof of Corollary 2.1, $u_t \partial_j [a_{jk}\partial_k \phi_m] \to 0$ in L^2 as $m \to \infty$. The fact that $Hu_t \in L^2$ requires some argument, which we shall not present here. However, since it is the case, we see that $\phi_m^2 u_t \to u_t$ in L^2 and from (3.11) we see that $H\phi_m^2 u_t \to Hu_t$ in L^2. Since H_{min} is closed, $u_t \in \mathcal{D}(H_{min})$. An application of Nelson's Lemma 3.1 shows that H_{min} is selfadjoint.

For an operator which is not potential free, the proof works approximately the same way. Consider the operator

$$(3.11) \qquad H^q = \sum_{j,k} \partial_j a_{jk} \partial_k - q, \qquad q \geq 0.$$

We have a corresponding semi-group given by the Kac-Feynman integral

$$(3.12) \qquad T_t^q\, u(x) = E\left[e^{-\int_0^t q(\xi_x(s))ds}\, u(\xi_x(t))\right].$$

It may be shown, in the standard way, by the use of Ito's formula for differentials, that the differential generator of the semi-group $\{T_t^q\}$ is H^q of (3.12). It also may be shown that this semi-group is a selfadjoint C_0 semi-group on $L^2 \to L^2$.

The proof we have outlined above requires very stringent conditions on the matrix $(a_{jk}(x))$, namely (3.5) and (3.6). By refining these techniques one is able to get the following considerably better theorem.

Theorem 3. Suppose the matrix $[a_{jk}(x)]$ is positive semi-definite, $a_{jk} \in C^2(R^n)$ with bounded second derivatives and $q \in C^1(R^n)$, $q \geq 0$, and

$$|\nabla q(x)| \leq K(1 + |x|^m)(1 + q(x)).$$

Then the minimal operator associated with H^q of (3.11) is selfadjoint.

The technique of proof consists in setting $a_{jk}^\varepsilon = a_{jk} + \varepsilon\delta_{jk}$, $\varepsilon > 0$. Then by suitably mollifying a_m^ε one can show that there exists a matrix σ^ε with the properties (3.4) and (3.5). The next step is to allow $\varepsilon \to 0$. Full details will appear elsewhere.

References

1. A. Devinatz, Essential selfadjointness of Schrödinger type operators, J. Functional Anal. (to appear).

2. ———, Essential selfadjointness of certain partial differential operators on R^n, Proc. Amer. Math. Soc. (to appear).

3. E. B. Dynkin, Markov Processes, Vol. I, Academic Press and Springer-Verlag, 1968.

4. M. S. P. Eastham, W. D. Evans and J. B. McLeod, The essential selfadjointness of Schrödinger-type operators, Arch. Rational Mech. Anal. 60 (1976) 185-204.

5. I. I. Gihman and A. V. Skorohod, Stochastic differential equations, Springer-Verlag, 1972.

6. T. Kato, Perturbation theory for linear operators, Springer-Verlag, 1966.

7. ———, Schrödinger operators with singular potentials, Israel J. Math. 13 (1972) 135-148.

8. ———, A second look at the essential self-adjointness of the Schrödinger operator, in Physical Reality and Mathematical Description, Eds. C. P. Enz and J. Mehra, D. Reidel Pub. Co., Dordrecht, 1974, pp. 193-201.

9. I. Knowles, On essential self-adjointness for singular elliptic differential operators, (to appear).

10. S. A. Laptev, Closure in the metric of a generalized Dirichlet integral, Differential Equations, vol. 7, no. 4 (1971) 727-736.

11. E. Nelson, Analytic vectors, Ann. Math., 70 (1959) 572-615.

12. N. N. Ural'ceva, The nonselfadjointness in $L_2(R^n)$ of an elliptic operator with rapidly increasing coefficients (Russian), Zap. Naucn. Sem. Leningrad, Otedl. Mat. Inst. Steklov (LOMI), 14 (1969) 288-294.

SELF-ADJOINT DIFFERENTIAL EQUATIONS WITH ALL SOLUTIONS $L^2(0,\infty)$

M. S. P. EASTHAM

1. Introduction

In the terminology of deficiency index theory, a differential equation

$$\sum_{k=0}^{n} \{p_k(x)y^{(k)}(x)\}^{(k)} = \lambda y(x) \quad (0 \leqslant x < \infty) \qquad (1.1)$$

is said to be of limit-N type if the equation has N linearly independent solutions which are $L^2(0,\infty)$. Here im $\lambda \neq 0$ and we impose the straightforward conditions that the real-valued coefficient $p_k(x)$ is $C^{(k)}[0,\infty)$ ($k = 0, 1, \ldots, n$) and that $p_n(x) > 0$ in $[0,\infty)$. The possible values for N are

$$n,\ n+1,\ \ldots,\ 2n.$$

We shall leave aside the familiar case $n = 1$ and concentrate on (1.1) with $n \geqslant 2$ and particularly $n = 2$.

Conditions on the $p_k(x)$ under which $N = n$ have been the subject of considerable recent investigations (see e.g. Atkinson (2) and Devinatz (5)). In contrast, very little is known about situations for which $n + 1 \leqslant N \leqslant 2n$. There are some results for $n = 2$, $N = 3$ in (6) but apart from this the only general method available for $n + 1 \leqslant N \leqslant 2n$ is the complicated one of determining the asymptotic form of the solutions of (1.1) as $x \to \infty$. For this method we refer to Devinatz (3), Fedorjuk (9), Kogan and Rofe-Beketov (12) and, for a fuller development in the case $n = 2$, to Devinatz (4) and Walker (15, 16). The work of Devinatz and Walker is striking on account of the complicated nature of the transformations which are required to bring the differential equation into a form to which the Levinson asymptotic theorem applies.

In this paper, I discuss a very much simpler method for investigating the case $N = 2n$. Although the method does not go so far as to determine the precise asymptotic form of solutions, it does give enough information to allow their L^2 nature to be decided, subject of course to certain conditions on the $p_k(x)$. An important simplification attaching to the case $N = 2n$ is that the right-hand side of (1.1) can be replaced by zero and we shall take advantage of this fact.

Our method is based on a modification of work of Kuptsov (14) for second-order systems. In §2, we give this modification and in §§3-4 we give applications to fourth-order equations. Then in §5 we consider the possibility of dealing with higher-order equations.

2. Second-order systems

We consider the second-order system

$$Y''(x) + P(x)Y'(x) + Q(x)Y(x) = 0, \qquad (2.1)$$

where P and Q are real nxn matrices and Y is a real n-component vector, and we write (2.1) as

$$Z'(x) = A(x)Z(x), \qquad (2.2)$$

where

$$A(x) = \begin{pmatrix} 0 & I \\ -Q(x) & -P(x) \end{pmatrix}, \quad Z(x) = \begin{pmatrix} Y(x) \\ Y'(x) \end{pmatrix}.$$

Let V(x) be an arbitrary self-adjoint positive 2n×2n matrix. Then, by (2.2),

$$\frac{d}{dx}(VZ,Z) = ((V' + VA + A^*V)Z,Z),$$

where the asterisk denotes the adjoint matrix. Hence, if we define

$$\mu(x) = \sup_z((V' + VA + A^*V)z,z)/(Vz,z), \qquad (2.3)$$

where the sup is taken over all real 2n-component vectors z, we obtain

$$(VZ,Z) \leqslant (\text{const.})\exp\left(\int^x \mu(t)\,dt\right). \qquad (2.4)$$

We now choose

$$V = F^*UF,$$

where

$$F = \begin{pmatrix} I & 0 \\ D & I \end{pmatrix}, \quad U = \begin{pmatrix} T & 0 \\ 0 & L \end{pmatrix}$$

and D, T, L are n×n matrices with T and L positive and self-adjoint. We define

$$m_1(x) = \sup_u\{(T'u,u)/(Tu,u)\} \qquad (2.5)$$

$$m_2(x) = \sup_v\{((L' - LP - P^*L)v,v)/(Lv,v)\} \qquad (2.6)$$

and

$$\phi(x) = \tfrac{1}{4}\{m_1(x) - m_2(x)\}. \qquad (2.7)$$

Then, with the choice $D = dg(\phi, \ldots, \phi)$, it was shown in (7), as a modification of the work of Kuptsov (14), that (2.3) and (2.4) yield

$$(TY,Y) \leqslant (\text{const.})\exp\left(\int^x \left[\tfrac{1}{2}\{m_1(t) + m_2(t)\} + |M(t)|\right] dt\right) \qquad (2.8)$$

for any solution Y of (2.1), where

$$M(x) = \sup_{u,v}\{(Gu,v)/(Tu,u)^{\frac{1}{2}}(Lv,v)^{\frac{1}{2}}\} \qquad (2.9)$$

and

$$G = L(D' - D^2 + PD) - LQ + T. \qquad (2.10)$$

The idea in using (2.8) is to arrange that M(t) is negligible in (2.8) and, to this end, the symmetric matrices L and T are to be chosen so that

$$LQ = T, \qquad (2.11)$$

as suggested by (2.10). Then, with $\phi(x)$ defined by (2.7), G is given by (2.10-11) and, under suitable conditions, $M(x)$ can be shown to be negligible in (2.8).

We conclude this section with some general remarks. The first is that Kuptsov's method and results (13, 14) appear to have been overlooked in the West until Knowles (11) drew attention to them recently. For example, Theorem 3.4 of (1) on the limit-circle case of the system $Y''(x) + P(x)Y(x) = 0$ is already covered by Kuptsov's paper (14). We also mention that the expression (VZ,Z) has a further application in the different context of the characteristic exponents for periodic systems (17, pp.134-9).

3. First application to fourth-order equations

Let $q(x)$ be a $C^{(2)}[0,\infty)$ function which increases to $+\infty$ as $x \to \infty$. Let

$$q'^2(x)q^{-5/2}(x) \text{ and } q''(x)q^{-3/2}(x) \text{ be } L(X,\infty) \tag{3.1}$$

for some X. Let a and b be positive constants and consider the fourth order differential equation

$$y^{(4)}(x) + a\{q(x)y'(x)\}' + \{bq^2(x) + q''(x)\}y(x) = 0. \tag{3.2}$$

In the particular case $a = 10/3$, $b = 1$, the solutions of (3.2) are $y = z_1z_2z_3$, where the z_i (not necessarily distinct) are solutions of $z'' + \frac{1}{3}qz = 0$. Since $z = O(q^{-\frac{1}{4}})$, we have $y = O(q^{-\frac{3}{4}})$ as $x \to \infty$ in this case. For (3.2) generally, we can prove that

$$y(x) = O\{q^{-k}(x)\}, \tag{3.3}$$

where

$$k = \tfrac{1}{4}[2 - \{(b^{-1} + a - 3)/(a - 1 - b)\}^{\frac{1}{2}}], \tag{3.4}$$

provided that

$$0 < b < a - 1. \tag{3.5}$$

The estimate (3.3) immediately gives a condition under which all solutions $y(x)$ of (3.2) are $L^2(0,\infty)$, that is, the equation (3.2) is of limit-4 type.

To prove (3.3) we write (3.2) as a system (2.1) with

$$Q = \begin{pmatrix} q & 1 \\ cq^2 & (a-1)q \end{pmatrix}, \quad P = \begin{pmatrix} 0 & 0 \\ (2-a)q' & 0 \end{pmatrix} \tag{3.6}$$

where $c = a - 1 - b$. A choice for L and T satisfying (2.11) is

$$L = \begin{pmatrix} (a-1)q & -1 \\ -1 & c^{-1}q^{-1} \end{pmatrix}, \quad T = \begin{pmatrix} bq^2 & 0 \\ 0 & c^{-1}b \end{pmatrix} \tag{3.7}$$

and (3.5) makes L and T positive. Omitting details of the calculation, which can be found in (7), we obtain from (2.5-6)

$$m_1(x) = 2q'(x)/q(x)$$

and $$m_2(x) = (2 - 4k)q'(x)/q(x).$$

Also, (3.1) implies that $M(x)$ is $L(X,\infty)$. Then (2.8) gives

$$q^2(x)y^2(x) + c^{-1}(qy + y'')^2(x) \leqslant (\text{const.})\{q(x)\}^{2-2k},$$

from which (3.3) follows.

4. Second application to fourth-order equations

We consider the equation

$$\{r(x)y''(x)\}'' + \{p(x)y'(x)\}' + q(x)y(x) = 0, \qquad (4.1)$$

where $$r(x) = x^{\gamma},\ p(x) = ax^{\alpha},\ q(x) = bx^{\beta}, \qquad (4.2)$$

a, b are positive constants and α, β, γ are non-negative constants. We prove that

$$y(x) = O(x^{-\frac{1}{4}(\alpha-\gamma)+\epsilon}) \qquad (4.3)$$

as $x \to \infty$, for any $\epsilon > 0$, provided that

$$\alpha - 2 < \beta < 2\alpha - \gamma. \qquad (4.4)$$

Consequently, all solutions of (4.1) are square-integrable over $(0,\infty)$ if, further to (4.4), $$\alpha > 2 + \gamma. \qquad (4.5)$$

To prove (4.3), we write (4.1) as a system (2.1) with

$$Q = \begin{pmatrix} 0 & 1/r \\ -q & p/r \end{pmatrix}, \quad P = \begin{pmatrix} 0 & 0 \\ -p' & 0 \end{pmatrix}.$$

A suitable choice for L and T satisfying (2.11) is

$$L = \begin{pmatrix} p - rq(1+h)/p & -1 \\ -1 & (1+h)/p \end{pmatrix}$$

$$T = \begin{pmatrix} q & -q(1+h)/p \\ -q(1+h)/p & h/r \end{pmatrix},$$

where $h(x)$ is arbitrary except that

$$h > (1+h)^2 rq/p \qquad (4.6)$$

to make L and T positive. We choose $h(x) = x^{-\delta}$, where, by (4.2) and (4.6), $$0 < \delta < 2\alpha - \beta - \gamma.$$

Here the right-hand part of (4.4) is required. Then, omitting details of the calculation, which can be found in (8), we obtain from (2.5-6)

$$m_1(x) \sim \beta x^{-1},\ m_2(x) \sim (\alpha - \delta)x^{-1}. \qquad (4.7)$$

Also, from (2.9), $$M(x) = O(x^{\frac{1}{2}(\alpha-\beta)-2}), \qquad (4.8)$$

and so $M(x)$ is $L(X,\infty)$ by the left-hand part of (4.4). Then (2.8) gives $$x^{\beta}y^2(x) \leqslant (\text{const.})x^{\frac{1}{2}(\alpha+\beta-\delta)},$$

and (4.3) follows on writing $\delta = 2\alpha - \beta - \gamma - 4\epsilon$.

The above method can also be applied when the inequalities in (4.4) are each replaced by equalities. For example, when $\beta = \alpha - 2$, (4.8) becomes $M(x) = O(x^{-1})$, and the constant implied by the O-notation ca be made explicit (8). Then $M(x)$ contributes to (2.8) along with $m_1(x)$ and $m_2(x)$ (see (4.7)). We find that all solutions of (4.1) are squar integrable over $(0, \infty)$ if

$$\beta = \alpha - 2, \quad \beta > \gamma,$$

and $$(b/a)^{\frac{1}{2}} > (3\beta^2 - 2\beta\gamma + 8\beta + 4 - \gamma^2)/\{8(\beta - \gamma)\}.$$

The more complicated asymptotic theory of Devinatz (4) and Walker (16) also requires (4.4) but (4.5) is replaced by the better conditi $\alpha + \beta > 2$. On the other hand, this theory does not deal with $\beta = \alpha - 2$. The work in (12) does cover $\beta = \alpha - 2$ but only if β is an integer ar $\gamma = \alpha$ as well, and then the condition on b/a is $b/a > \frac{1}{2}\beta + \frac{1}{4}$.

Finally in this section, we remark that it is not intended that th basic condition $r(x) > 0$ in $[0, \infty)$ be violated by (4.2). Our method is applicable to (4.1) if r, p, q merely resemble powers of x suffic iently closely as $x \to \infty$.

5. Higher-order equations

The general equation (1.1) can be written as a system (2.1) with suitable P and Q, the form of P and Q differing slightly according t whether n is even or odd. Also, symmetric L and T satisfying (2.11) exist, although the explicit determination of L and T rapidly become tedious with increasing n. The estimation of $m_1(x)$ and $m_2(x)$ is not as simple as in the fourth-order case but we can give some details here for n = 3 (the sixth-order case). We consider (1.1) with n = 3,

$$p_3(x) = x^\alpha, \quad p_2(x) = x^\beta, \quad p_1(x) = x^\gamma, \quad p_0(x) = x^\delta,$$

where $\alpha \geqslant 0$, $\beta > 0$, $\gamma > 0$, $\delta > 0$. Here Q, P, L, T are respectively

$$\begin{pmatrix} 0 & -1 & 0 \\ 0 & p_2/p_3 & -1/p_3 \\ p_0 & p_1 & 0 \end{pmatrix}, \quad \begin{pmatrix} 0 & 0 & 0 \\ 0 & p_3'/p_3 & 0 \\ p_1' & 0 & 0 \end{pmatrix},$$

$$\begin{pmatrix} p_1 + gp_0 - hp_2p_0 & -hp_3p_0 & 1 \\ -hp_3p_0 & p_3 + gp_2 - hp_3p_1 & -g \\ 1 & -g & h \end{pmatrix},$$

$$\begin{pmatrix} p_0 & -gp_0 & hp_0 \\ -gp_0 & R & S \\ hp_0 & S & g/p_3 \end{pmatrix},$$

where
$$R = p_2 - g(p_1 - p_2^2/p_3) + h(p_3p_0 - p_2p_1),$$
$$S = -1 - gp_2/p_3 + hp_1,$$
and g and h are arbitrary functions. The choice

$$g = (1 + c)p_3/p_2, \; h = (1 + c)/p_1$$

is made, where c is a positive constant. Under the conditions

$$\alpha - \beta < \beta - \gamma < \gamma - \delta, \tag{5.1}$$

a calculation gives

$$m_1(x) = x^{-1}\{1 + o(1)\}\max\{\delta, \beta(1 + c^{-1})^{\frac{1}{2}}\}, \tag{5.2}$$
$$m_2(x) = x^{-1}\{1 + o(1)\}\gamma(1 + c)^{\frac{1}{2}}, \tag{5.3}$$
$$M(x) = O(x^{-2+\frac{1}{2}(\gamma-\delta)}) \tag{5.4}$$

as $x \to \infty$. Thus, if in addition to (5.1)

$$\gamma - \delta < 2, \tag{5.5}$$

$M(x)$ is $L(X,\infty)$.

Our analysis gives no result about $L^2(0,\infty)$ solutions if $\delta \leqslant \beta$. We consider therefore
$$\delta > \beta \tag{5.6}$$
and in (5.2) we choose c to make

$$\beta(1 + c^{-1})^{\frac{1}{2}} \geqslant \delta. \tag{5.7}$$

Then, by (2.8) and (5.2-3), we find that all solutions $y(x)$ are square integrable over $(0,\infty)$ if, in addition to (5.1), (5.5), (5.6) and (5.7), we have
$$\beta(1 + c^{-1})^{\frac{1}{2}} + \gamma(1 + c)^{\frac{1}{2}} < 2\delta - 2. \tag{5.8}$$

We can write the various conditions on α, β, γ, δ in a form that can be more readily appreciated as follows. For some c with

$$0 < c < 9/16, \tag{5.9}$$

let
$$\delta > 4\{2 - c^{\frac{1}{2}} - (1 + c)^{\frac{1}{2}}\}^{-1} \tag{5.10}$$
$$\delta \leqslant (1 + c^{-1})^{\frac{1}{2}}\beta < (2 + c^{\frac{1}{2}})^{-1}[\delta\{4 - (1 + c)^{\frac{1}{2}}\} - 4] \tag{5.11}$$
$$\tfrac{1}{2}(\beta + \delta) < \gamma < (1 + c)^{-\frac{1}{2}}(2\delta - 2) - c^{-\frac{1}{2}}\beta \tag{5.12}$$
$$\alpha < 2\beta - \gamma. \tag{5.13}$$

Here, (5.12) arises from (5.1) and (5.8); (5.11) arises from (5.7) and the need to make (5.12) meaningful; (5.10) makes (5.11) meaningful; (5.9) makes
$$c^{\frac{1}{2}} + (1 + c)^{\frac{1}{2}} < 2,$$

as required in (5.10); (5.13) arises from (5.1). To check (5.6), we note that (5.9) and the right-hand part of (5.11) give

$$5\beta/3 < \tfrac{1}{2}(3\delta - 4) < 3\delta/2,$$

whence $$\beta < 9\delta/10 < \delta.$$

To check (5.5), we note that the left-hand part of (5.11) and the right-hand part of (5.12) give

$$\gamma < (1 + c)^{-\frac{1}{2}}(\delta - 2) < \delta - 2 < \delta + 2,$$

as required. In the same way, in terms of γ and β, we have

$$\gamma < c^{-\frac{1}{2}}\beta - 2(1 + c)^{-\frac{1}{2}} < 2\beta \qquad (5.14)$$

certainly if $c \geqslant \frac{1}{4}$. Then (5.13) allows positive α. More generally, (5.14) holds for a suitable range of β consistent with (5.10-11) if

$$c > 1/8. \qquad (5.15)$$

We note that (5.15) and (5.10) imply that $\delta > 2(2 + \sqrt{2})$, and therefore (5.10-12) imply that β, γ, and δ are relatively large.

As definite limit-6 examples in which all the above conditions are satisfied, we cite $c = \frac{1}{3}$ and

$$\delta = 20, \ \beta = 10, \ 15 < \gamma < 9\sqrt{3} \ (= 15.59), \ \alpha < 20 - \gamma;$$

also $c = 9/64$ and

$$\delta = 11, \ \beta = 4, \ 7.5 < \gamma < 8, \ \alpha < 8 - \gamma.$$

We note that these examples are not covered by the asymptotic theory of Fedorjuk (9).

In our choice of g and h (just preceding (5.1)), we took the same constant c in each. There appears to be no advantage to be gained by choosing different constants c for g and h. We also comment on the complexity of the conditions (5.9-13) as compared to those in the fourth-order case. If indeed these are the sort of conditions to be expected, then we have an explanation of why the more intricate and precise asymptotic analysis of (4) and (15, 16) has not proceeded beyond the fourth-order case.

It seems likely that (1.1) with general n can be handled by our methods provided that the corresponding symmetric matrices L and T can be written in a manageable form. In the case where

$$p_k(x) = x^{\alpha_k} \quad (x \geqslant X),$$

where $\alpha_k \geqslant 0$, the conditions

$$\alpha_n - \alpha_{n-1} < \alpha_{n-1} - \alpha_{n-2} < \cdots < \alpha_1 - \alpha_0 < 2 \qquad (5.16)$$

allow L and T to be determined with sufficient accuracy to let the method proceed, and $M(x)$ can be shown to be $L(X,\infty)$. The conditions (5.16) generalize (5.1) and (5.5). These details for general n are still being worked out and it is hoped that they will appear in due course.

Other matters under consideration (10) are the questions of

(i) determining $n\times n$ matrices Q and P, generalizing (3.6), which yield estimates $O\{q^{-k}(x)\}$ for solutions of (2.1),

(ii) extending the methods in §§3-5 to differential equations with complex-valued coefficients.

References

1. R.L. Anderson, Limit-point and limit-circle criteria for a class of singular symmetric differential operators, Canadian J. Math. 28 (1976) 905-14.

2. F.V. Atkinson, Limit-n criteria of integral type, Proc. Roy. Soc. Edinburgh (A) 73 (1975) 167-98.

3. A. Devinatz, The deficiency index of a certain class of ordinary self-adjoint differential operators, Advances in Math. 8 (1972) 434-73.

4. A. Devinatz, The deficiency index of certain fourth-order ordinary self-adjoint differential operators, Quart. J. Math. (Oxford) (2) 23 (1972) 267-86.

5. A. Devinatz, On limit-2 fourth-order differential operators, J. London Math. Soc. (2) 7 (1973) 135-46.

6. M.S.P. Eastham, The limit-3 case of self-adjoint differential expressions of the fourth-order with oscillating coefficients, J. London Math. Soc. (2) 8 (1974) 427-37.

7. M.S.P. Eastham, Square-integrable solutions of the differential equation $y^{(4)} + a(qy')' + (bq^2 + q'')y = 0$, Nieuw Arc. v. Wisk. (3) 24 (1976) 256-69.

8. M.S.P. Eastham, The limit-4 case of fourth-order self-adjoint differential equations, Proc. Roy. Soc. Edinburgh (A), to appear.

9. M.V. Fedorjuk, Asymptotic methods in the theory of one-dimensional singular differential operators, Trans. Moscow Math. Soc. 15 (1966) 333-86 in the A.M.S. translation.

10. S.B. Hadid, London University Ph.D. thesis, in preparation.

11. I. Knowles, On second-order operators of limit-circle type, Proc. Dundee Conference, Lecture Notes in Mathematics 415 (Springer, 1974) 184-7.

12. V.I. Kogan and F.S. Rofe-Beketov, On the question of the deficiency indices of differential operators with complex coefficients, Proc. Roy. Soc. Edinburgh (A) 72 (1973/4) 281-98.

13. N.P. Kuptsov, Conditions of non-self-adjointness of a second-order linear differential operator, Dokl. Akad. Nauk. 138 (1961) 767-70.

14. N.P. Kuptsov, An estimate for solutions of a system of linear differential equations, Uspehki Mat. Nauk 18 (1) (1963) 159-64.

5. P.W. Walker, Deficiency indices of fourth-order singular differential operators, J. Diff. Equations 9 (1971) 133-40.

6. P.W. Walker, Asymptotics for a class of fourth-order differential equations, J. Diff. Equations 11 (1972) 321-34.

7. V.A. Yakubovich and V.M. Starzhinskii, Linear differential equations with periodic coefficients, Vol.1 (Wiley, I.P.S.T., 1975).

Chelsea College (University of London)
London SW3 6LX.

On the deficiency index problem for ordinary differential operators 1910-1976

W N Everitt

Introduction

This paper surveys, albeit briefly, the main results in the theory of deficien indices for ordinary symmetric differential expressions starting from the now classic memoir of Hermann Weyl which appeared in 1910. Neither the paper nor the list of references has any pretensions to completeness, space available will not support such a venture on this occasion, but it is hoped that a detailed survey of results in this area will appear in due course in Nieuw Archief voor Wiskunde. However it is hoped that all major sources of reference have been included here; other works not explicitly represented may be found in the references of the books and papers here listed.

The deficiency index problem stradles both classical and functional analysis; some results are best obtained by classical methods whilst others have their most elegant proofs in operator theoretic methods. Whilst most books employ both method it is not without interest to remark of the books quoted in the references, that those of Atkinson [1], Hille [26], Titchmarsh [55], Yosida [60] concentrate mainly on the classical side and those of Dunford and Schwartz [9], Levitan and Sargsjan [40], Naimark [42] and Stone [54] on operator theoretic techniques.

No explicit mention is made of the contribution of J. von Neumann since he was not primarily concerned with the deficiency index problem at any stage. However his outstanding contributions to the general theory of linear operators in Hilbert space had a more than significant effect on the theory of differential operators and reference to his work may be found in the bibliography of [54]. A more direct contribution from von Neumann may be seen in the work of Halperin [23].

No attempt is made to give any formal statement of results or proofs in this paper. However it is hoped that the many references to books and papers will enabl the reader to seek details as interest demands.

The paper is in three sections. Section 1 refers to the original memoir [57] of Weyl which appeared in 1910. Section 2 covers the range and coefficient problem for an important class of symmetric differential expressions and their associated deficiency indices; Section 3 is concerned with ramifications and extensions of the deficiency index problem.

Section 1

The first sign of the deficiency index problem for ordinary, symmetric differential equations appeared in the fundamental paper of Weyl [57] in 1910. (The use of the adjective symmetric here is described in some detail in Section 2; this replaces the earlier term ' formally self-adjoint' which is inappropriate in a sense also to be made clear in Section 2.)

The memoir [57] is concerned with properties of the following linear, ordinary differential equation of the second-order

$$-\left(p(x)y'(x)\right)' + q(x)y(x) = \lambda y(x) \qquad (x \in [0,\infty))$$

or

$$-(py')' + qy = \lambda y \quad \text{on} \quad [0,\infty). \tag{1.1}$$

Here p, q are real-valued coefficients on the half-line $[0,\infty)$ satisfying

(i) p and q are continuous on $[0,\infty)$

(ii) $p(x) > 0$ for all $x \in [0,\infty)$;

λ is a complex-valued parameter and ' denotes differentiation with respect to the independent variable x.

The conditions on the coefficients p and q ensure that the equation (1.1) has global solutions on $[0,\infty)$ determined by initial conditions at 0. As the notation used in [57] is not consistent with more modern notations we use the notation of [55]; thus if the solutions θ and ϕ of (1.1) are determined by the initial conditions (we write $\theta(x,\lambda)$ for the solution to indicate its dependence on the variable x and the parameter λ)

$$\theta(0,\lambda) = 1 \qquad p(0)\theta'(0,\lambda) = 0 \tag{1.2}$$

$$\phi(0,\lambda) = 0 \qquad p(0)\phi'(0,\lambda) = 1 \tag{1.3}$$

for all $\lambda \in C$ (the complex plane), then $\theta(\cdot,\lambda)$ and $\phi(\cdot,\lambda)$ satisfy (1.1) on $[0,\infty)$ and form a linearly independent basis for all solutions of the equation. For all $x \in [0,\infty)$ the functions $\theta(x,\cdot)$, $\phi(x,\cdot)$, $p(x)\theta'(x,\cdot)$ and $p(x)\phi'(x,\cdot)$ are integral (entire) analytic functions on C.

The proof of these statements given by Weyl, see [57, Kapitel I, Section 1] depends upon obtaining a uniformly and absolutely convergent series expansion of the solution; in the case of the solution θ, determined by (1.2), this takes the form, where $Q(t,\lambda) = q(t) - \lambda$,

$$\theta(x,\lambda) = 1 + \sum_{n=1}^{\infty} \iint \cdots \iint_{0\le\tau_1\le t_1\le\cdots\le\tau_n\le t_n\le x} \frac{Q(\tau_1,\lambda)\cdots Q(\tau_n,\lambda)}{p(t_1)\cdots p(t_n)}\, d\tau_1\, dt_1 \cdots d\tau_n\, dt_n . \tag{1.4}$$

It is interesting to note that since no differentialability requirement is placed on the coefficient p this existence theorem shows that a solution y of (1.1) has a continuous derivative y' on $[0,\infty)$ and (1.1) is satisfied at all points of the half-line. In general the conditions do not allow of the separate existence of the second derivative y". Here Weyl anticipated the introduction of the idea of quasi-derivatives, see [42, Section 15.2], and we comment on this point again in Section 2 below.

With the framework set we quote now from the opening remarks made by Weyl, see [57, Page 220],

"In the first part of this work the Green's function is defined which makes it possible to replace the differential equation by an integral equation, and at the same time introduce the classification, which is fundamental to all that follows, of the differential equation as one of two types, the limit-point and limit-circle cases."

In these words was issued in one of of the significant problems in analysis; the search for necessary and sufficient conditions on the coefficients p and q to distinguish between these two cases. One mark of distinction of greatness in a mathematician is the ability to select and work on the significant problems in mathematics and in this way, as with so many other problems did Hermann Weyl distinguish himself. To quote, from another context, the Scottish-American mathetician and historian E. T. Bell "the scores of profound papers bristling with thorny analysis" which have since been written on the limit-point/limit-circle classificati problem bear witness to the importance of Weyl's early work in ordinary differential equations and to the significant ramifications of these problems.

To introduce the limit-point/limit-circle classification we quote from [57, Page 227]

Theorem 2 The differential equation

$$-(py')' + qy = \lambda y \quad \text{on} \quad [0,\infty) \tag{1.1}$$

when λ is not real, always has a solution (not the null solution) which is absolutel square-integrable on the interval $[0,\infty)$.

This remarkable result (no additional constraints are required on the coefficients p and q) is best seen using again the notation of Titchmarsh [55, Chapter II]. With θ and ϕ determined by the initial conditions (1.2) and (1.3) the essence of the result in this Theorem is the existence of a coefficient $m : C \to C$, regular (Cauchy analytic) in the two half-planes of C created by the real line, such that the solution

$$\psi(x,\lambda) = \theta(x,\lambda) + m(\lambda)\phi(x,\lambda) \quad (x \in [0,\infty))$$

satisfies

$$\int_0^\infty |\psi(x,\lambda)|^2 dx < \infty \qquad (\lambda \in C \text{ with im } \lambda \neq 0).$$

It follows that at any point $\lambda \in C$ with im $\lambda \neq 0$

either (i) both $\theta(\cdot,\lambda)$ and $\phi(\cdot,\lambda)$ are not integrable-square

(1.5)

or (ii) both $\theta(\cdot,\lambda)$ and $\phi(\cdot,\lambda)$ are integrable-square.

For reasons concerned with the method of proof of [57, Theorem 2] quoted above, Weyl called the first of these two alternatives the limit-point case, and the second alternative the limit-circle case.

The second significant result in the classification of the differential equation (1.1) we quote from [57, Page 238]

Theorem 5 In the limit-circle case, for each λ (real or complex) the differential equation

$$-(py')' + qy = \lambda y \quad \text{on} \quad [0,\infty) \tag{1.1}$$

has only absolutely square-integrable solutions; in the limit-point case, on the other hand, there is not a single value of λ (real or complex) for which this equation has two such linearly independent solutions.

In such terms Weyl expressed the fact that the limit-point/limit-circle classification of the differential equation is independent of the parameter λ. In other words if (i) or (ii) of (1.5) hold at one point λ then the same classification holds at all strictly complex points of C. In the limit-circle case all solutions are absolutely square-integrable also for all real values of λ. In the limit-point case $\theta(\cdot,\lambda)$ or $\phi(\cdot,\lambda)$ may be absolutely square-integrable for certain real values of λ but not both at the same point.

We shall relate these results to the corresponding results in the theory of deficiency indices in the next Section.

It is clear from these Theorems of [57] that the limit-point/limit-circle classification of the differential equation (1.1) depends only on the coefficients p and q. Although not stated explicitly in [57] this leads to the central problem in the classification of the differential equation (1.1)

Problem To find necessary and sufficient conditions on the coefficients p and q in order to determine if the equation (1.1) is in the limit-point, or the limit-circle, case.

In spite of the most determined efforts over the years since 1910 there is as yet no definite solution to this problem although there is now a considerable, indeed formidable, and growing corpus of sufficiency conditions to determine either that the limit-point or the limit-circle condition holds.

The first such condition was given by Weyl himself as a Corollary to Theorem 5 of [57]; if for some real number c the coefficient q satisfies $q(x) \geq c$ for all $x \in [0,\infty)$ then the differential equation is in the limit-point case. The proof of this result follows at once from the representation (1.4) for the solution $\theta(\cdot,\lambda)$; if in (1.4) we take $\lambda = c$ then all terms of the infinite series are seen to be non-negative and so

$$\theta(x,c) \geq 1 \text{ for all } x \in [0,\infty);$$

clearly this implies that $\theta(\cdot,c)$ is not absolutely square-integrable on $[0,\infty)$ and that the limit-point case holds.

From the classical viewpoint of the theory of differential equations there are now several accounts of the above mentioned general theory and the limit-point/limit -circle classification; see the book [26, Chapter 10], [55, Chapters II and V] and [60, Chapter 5]; for an alternative form of proof of these results see Everitt [15] and [16].

For a part translation and a commentary on the memoir [57] of Weyl see the MSc thesis of Race [47].

For significant results in the limit-point/limit-circle classification problem see Hartman [24], Hartman and Wintner [25], Ismagilov [28], Knowles [33], Levinson [39], Sears [50] and [51], Wong [59]. For a survey of nearly all results up to 1972 see the PhD thesis of Knowles [34]. The paper by Evans [13] contains a number of recent and very comprehensive results; likewise the paper by Atkinson [2]. The survey article by Giertz [21a] contains much valuable information on recent results. The recent paper of Kwong [38] solved an outstanding perturbation problem in the limit-point/limit-circle classification theory, but see initially the very interesting remarks by Wong [59].

Some idea of the very delicate balance between the limit-point and limit-circle conditions of the differential equation(1.1) may be gathered from the very remarkable examples constructed in the paper by Eastham and Thompson [10].

Many new ideas in the limit-point/limit-circle theory have been introduced in recent years and a brief account, with reference to many of the above mentioned papers, is given in the paper by Everitt [18].

The example discussed by McLeod [41] shows that if the differential equation is written in the form (1.1) then it is essential to take the coefficients p and q as real-valued if the general results of Weyl in [57] are to hold.

Finally reference should be made to the comprehensive account in the book [9, Chapter XIII]; see in particular the historical account given in [9, Chapter XIII, Section 10A].

It is to be stressed again that the references given above form only a part, and a minor part at that, of the body of results in this area.

Towards the end of his long life of labour in the vineyard of mathematics Weyl looked back on his early work in differential equations in his address to the American Mathematical Society "Ramifications, old and new, of the eigenvalue problem", see [58, Section 2]. In considering again the limit-point/limit-circle classification problem he wrote, with notable modesty, "The very first result by which I added my mite to our stock of mathematical knowledge had to do with the clarification of this issue". In viewing the scene ourselves after the passage of more than 60 years we may consider ourselves justified in releasing Hermann Weyl from his gentle modesty and replacing the word 'mite' with 'might'.

Section 2

The extension of the ideas considered in Section 1 rests with the introduction of general linear differential expressions of arbitrary integral order with real or complex coefficients. If we denote such a differential expression by

$$M[y](x) \stackrel{\text{def}}{=\!=} a_n(x)y^{(n)} + a_{n-1}y^{(n-1)}(x) + \ldots + a_0(x)y(x), \tag{2.1}$$

where the coefficients a_r $(r = 0,1,\ldots,n)$ are in general complex-valued, then the differential equation to be considered is

$$M[y] = \lambda\, w\, y \quad \text{on} \quad I \tag{2.2}$$

where, as before, λ is a complex parameter; I is an interval (bounded or unbounded; open, closed or half-open/half-closed) of the real line R; w is a positive (Lebesgue almost everywhere on I) 'weight' function. The general problem then is to search for solutions of (2.2) which lie in the integrable-square function space $L_w^2(I)$, i.e. the space of complex-valued functions determined by

$$\int_I w|f|^2 = \int_I w(x)|f(x)|^2 dx < \infty. \tag{2.3}$$

To link with the original problem of Weyl considered in Section 1 we take $I = [0,\infty)$ and

$$M[y] = -(py')' + qy\,, \quad w(x) = 1 \qquad (x \in [0,\infty)). \tag{2.4}$$

For a general differential expressions M it turns out that the problems posed by the equation (2.2) in respect of solutions in $L_w^2(I)$ are too general to yield a structure of results similar to those given in Section 1. Examples may readily be constructed to show that the extension of such results is impossible in general.

The solution to this situation lies in the introduction of a restriction on M which in turn yields a structure on the solutions of the equation (2.2) in $L_w^2(I)$, reducing to the results of Weyl in the special case (2.4). This restriction is one of symmetry on M.

A differential expression M as given by (2.1) is said to be symmetric (in an older and, in part, misleading terminology - formally self-adjoint) if $M = M^+$ where the adjoint differential expression M^+ is defined by

$$M^+[y] = (-1)^n(a_n y)^{(n)} + (-1)^{n-1}(\bar{a}_{n-1}y)^{(n-1)} + \dots + \bar{a}_0 y. \qquad (2.5)$$

This definition requires certain differentiability properties on the coefficients a_r $(r = 0,1, \dots, n)$.

For a symmetric differential expression M we have Green's formula, valid for functions f and g which are suitably differentiable on I and for all compact sub-intervals $[\alpha,\beta]$ of I,

$$\int_\alpha^\beta \{\bar{g}M[f] - f\overline{M[g]}\} = [fg](\beta) - [fg](\alpha). \qquad (2.6)$$

Here $[fg](\cdot)$ is a sesqui-linear form in f and g and their first n - 1 derivatives on I.

Various extensions of the definition of symmetry for differential expressions have been given which avoid the restrictive differentiability requirements of (2.5 Details of these extensions are too technically involved to be given here but we mention the main sources of references. The main advance was made by the Russian mathematician Shin who introduced the idea of quasi-derivatives in 1938; see [52] and the later paper [53]. The ideas of Shin were in part previously considered by Halperin in his paper [23] already mentioned. These concepts were taken up in the real even-order case of M particularly by Naimark, see the book [42] and the references therein, and Glazman in his memoir [22]. Later work in this direction may be found in the book by Atkinson [1], the paper by Walker [56] but in particular in the work of Zettl reported on in [61]. Zettl rediscovered and extended the work of Shin; the paper [61] is dated 1975 but much of the work involved dates from the year 1965. It should be mentioned here that the Shin-Zettl definition of symmetry permits the formal multiplication of these general differential expressions without the assumption of smoothness conditions on the coefficients; for some details see [62].

These results permit the definition of very general symmetric differential expressions for which the Green's formula (2.6) is available, and this now appears as a fundamental and indispensable tool in the development of the theory of Weyl.

Roughly speaking this development proceeds in two different but interwoven ways. The first uses methods essentially classical analysis and studies properties of the differential equation (2.2). The second uses operator theoretic methods in the Hilbert function space of equivalence classes in $L^2_w(0,\infty)$. The first procedure is the essential inheritance of the original methods of Hermann Weyl in [57], whilst the second used methods not available in 1910 but now frequently employed. Some contributions are difficult to classify in this respect and we make no attempt here to discern fine distinctions; rather the account flits in and out of both ways. However it is helpful to classify, as far as possible, the contributions listed in the references and then to present the results without specific reference to author or method, except for certain important results given below.

On the differential equation method contributions have been made by Atkinson [1], Everitt [14], [16] and [17], Everitt and Kumar [20] and [21], Kimura and Takahasi [32], Kodaira [35], Kogan and Rofe-Beketov [36] and [37], Nitsche [44], Pleijel [45] and [46], Rofe-Beketov [49], Titchmarsh [55], Walker [56].

For operator theoretic methods see Coddington [5] and [6], Dunford and Schwartz [9], Glazman [22], Halperin [23], Naimark [42], Stone [54].

An important survey article on certain aspects of this work was written by Devinatz [8].

We now suppose given a differential expression M of order n on an interval I of R, which is symmetric in the classical or generalized sense mentioned above, and which is regular on I, i.e. the differential equation (2.2) can be solved for any initial conditions at any point of I and such solutions exist globally on I.

In the Hilbert function space $L^2_w(I)$ two fundamental differential operators are defined, T_{min} and T_{max}; both operators are determined in the form

$$T_{min} f = w^{-1}M[f] \qquad T_{max} f = w^{-1}M[f]$$

for f in the corresponding domains $D(T_{min}) \subset D(T_{max})$. (Here w^{-1} denotes the reciprocal function $w^{-1}(x) = \{w(x)\}^{-1}$ $(x \in I)$ and not an inverse function). T_{min} is the smallest, closed symmetric operator which can be generated by $w^{-1}M$ in this way; T_{max} is the largest closed operator similarly associated with $w^{-1}M$.

These two operators are closely connected and satisfy the relationship (here * denotes an adjoint operator)

$$T_{min}^* = T_{max} \qquad T_{max}^* = T_{min} \tag{2.7}$$

The Green's formula (2.6) is essential to proving this result.

The deficiency $N(\lambda)$ of T_{min} at any point λ of the complex plane C is the cardinal number defined by

$$N(\lambda) = \dim \{f \in D(T_{min}^*) \mid T_{min}^* f = \lambda f\}$$

$$= \dim \{f \in D(T_{max}) \mid T_{max} f = \lambda f\}, \quad (\text{from } (2.7)).$$

Thus $N(\lambda)$ is finite or countable since $L_w^2(I)$ is a separable Hilbert space.

It is a general property of symmetric operators that $N(\lambda)$ is independent of λ in $C_+ = \{\lambda \in C \mid im[\lambda] > 0\}$, and also in $C_- = \{\lambda \in C \mid im[\lambda] < 0\}$, although these two numbers may be different. We now define the deficiency indices of T_{min} (equivalently of the differential expression $w^{-1}M$ on I) as the pair (N_+, N_-) of cardinal numbers given by

$$N_+ = N(i) \qquad N_- = N(-i). \tag{2.8}$$

The property (2.7) and a knowledge of the nature of the domain $D(T_{max})$ enables us to show that

$$N(\lambda) = \dim\{y : I \to C \mid y \text{ has n quasi-derivatives},$$
$$M[y] = \lambda wy \text{ on } I, \quad y \in L_w^2(I)\} \tag{2.9}.$$

It is this last property which links the deficiency indices of T_{min} with the number of linearly independent solutions of $M[y] = \lambda wy$ on I which also lie in $L_w^2(I)$, and makes possible the interplay between the differential equation and operator theoretic methods.

Note also that the number of such solutions of the differential equation (2.2) which are in $L_w^2(I)$ is independent of λ in C_+, and also in C_-; on fact these numbers are exactly N_+ and N_-.

Another immediate consequence of (2.9) is

$$0 \leq N_+, \quad N_- \leq n. \tag{2.10}$$

Thus all closed, symmetric extensions of T_{min} are finite dimensional extension; from the general theory of operators in Hilbert space T_{min} has self-adjoint extensions if $N_+ = N_-$, and T_{min} is self-adjoint if and only $N_+ = N_- = 0$, in which case $T_{min} = T_{max}$.

In addition to the constraints on N_+ and N_- given by (2.10) it may also be shown using operator theoretic or differential equation methods that

$$\underline{\text{for all}}\ n > 1 \quad N_+ = n\ \underline{\text{if and only if}}\ N_- = n. \tag{2.11}$$

This result is essentially false when n = 1.

With these results given we may state the so-called range problem for the deficiency indices of ordinary differential operators; this consists of determining all possible values of N_+ and N_- on being given the interval I and the order n; with I and n fixed the deficiency indices vary with the choice of coefficients which determine the differential expression $w^{-1}M$. This problem was essentially solved for real (i.e. real-valued coefficients) even-order symmetric differential expressions in 1950 but the problem still remains open in the complex, general order case.

We can simplify the framework of this problem by classifying the nature of the interval I (recall that $w^{-1}M$ is of order n and regular at all points of I)

(i) suppose $I = [a,b]$ is compact; then all solutions of the equation are in $L^2_w(a,b)$ and

$$N_+ = N_- = n$$

(ii) suppose $I = (a,b)$ is open, i.e. $-\infty \leq a < b \leq \infty$; let $c \in (a,b)$ and consider $w^{-1}M$ on the sub-intervals $(a,c]$ and $[c,b)$; let $N_\pm$, $L_\pm$ and $R_\pm$ be the deficiency indices of $w^{-1}M$ on (a,b), $(a,c]$ and $[c,b)$ respectively; then

$$N_\pm = L_\pm + R_\pm - n; \tag{2.12}$$

this result informs us that we need only consider in detail the cases when I is of the form $[a,b)$ or $(a,b]$

(iii) suppose $I = [a,b)$; i.e. $-\infty < a < b \leq \infty$, (the results given are the same for $I = (a,b]$) then the general bounds (2.10) may be replaced by

when n is even, $n = 2r$ $(r = 1,2,\ldots)$

$$r = \frac{1}{2}n \leq N_+, \quad N_- \leq n = 2r \tag{2.13}$$

when n is odd, $n = 2r - 1$ $(r = 1,2,\ldots)$

either $r - 1 \leq N_+ \leq n = 2r - 1$ and $r \leq N_- \leq n = 2r - 1$ (2.14)

or $r \leq N_+ \leq n = 2r - 1$ and $r - 1 \leq N_- \leq n = 2r - 1$ (2.15)

(in the case when n is odd the either/or situation is dependent on the choice of certain coefficients in $w^{-1}M$).

The results given in (i), (ii) and (iii) above represent the best possible bounds in each case for I; however it is still not clear whether all possibilities of N_+ and N_- subject to these bounds and (2.11) can be realised by choice of suitable differential expressions $w^{-1}M$; we list below what is known at present in the first six cases $n = 1,2,\ldots,6$ giving all the possibilities in each case and then remarking on those possibilities not yet realized; it should be noted that these are concerned with $I = [a,b)$;

n = 1 (0,1) (1,0) (1,1); all realized

n = 2 (1,1) (2,2); both realized

(see the remarks below on the relationship between this result and the results of Weyl quoted in Section 1)

n = 3 (1,2) (2,1) (2,2) (3,3); all realized

n = 4 (2,2) (2,3) (3,2) (3,3) (4,4); all realized

<u>n = 5</u> (2,3) (3,2) (2,4) (4,2) (3,3) (3,4) (4,3) (4,4) (5,5);
all realized except the two possibilities (2,4) and (4,2)

<u>n = 6</u> (3,3) (3,4) (4,3) (3,5) (5,3) (4,4) (4,5) (5,4) (5,5) (6,6)
all realized except the two possibilities (3,5) and (5,3)

In general, for all $n \geq 2$, it is known that all possibilities can be realized subject to the constraints (2.11), one of (2.13), (2.14), (2.15), whichever one is applicable, and the additional constraint

$$|N_+ - N_-| \leq 1. \qquad (2.1$$

It is this last constraint which makes the cases (2,4) for $n = 5$, and (3,5) for $n = 6$ difficult to realize. However it is now known that (2.16) is not a natural feature of the problem; quite recently Gilbert [21b] has shown, for the case $n = 7$ that it is possible to have $|N_+ - N_-| = 2$ and other results in this direction are likely to follow. However the cases $n = 5$ and $n = 6$ are still not settled and remain as above.

These recent results consolidate the conjecture that every possibility should be realizable, when I has the form [a,b), subject only to the constraints (2.11) a the relevant one of (2.13), (2.14), (2.15). If this is indeed so then when I has the form (a,b) every possibility for (N_+, N_-) will be realizable subject only to the constraints (2.10) and (2.11).

One additional comment should be made. At the Uppsala 1977 Conference of which these Proceedings form a record, R C Gilbert announced that it is possible to construct differential expressions of order n, defined on the half-line $[0,\infty)$, so that given any positive integer $p \geq 2$ the deficiencies indices (N_+, N_-) satisfy

$$|N_+ - N_-| = p.$$

For this result to hold it is necessary to choose the order $n = 4p - 1$, <u>e.g</u>. if $p = 2$ then $n = 7$ (see the remarks following (2.16)). Details of this result will be published in due course.

It is convenient at this point to return to the results of Weyl from [57] and quoted in Section 1 above. Theorem 2 from [57] is equivalent to the result given above that when $n = 2$ and I is of the form [a,b) then $1 \leq N_+, N_- \leq 2$, <u>i.e</u>. there is at least one integrable-square solution when the parameter λ is strictly complex; indeed the limit-point and limit-circle cases are precisely equivalent to the deficiency cases (1,1) and (2,2). On the other hand Theorem 5 from [57] is equivalent to the invariance of the deficiency $N(\lambda)$ of the operator T_{min} in the half-spaces C_+ and C_-, and to the 'maximal' property (2.11), <u>i.e</u>. $N_+ = 2$ if and only if $N_- = 2$, in the second-order case. The real and permanent significance of Weyl's results in this area may now be seen in the context of this wider framework the great mathematicians have an unerring way of settling upon the important results even though the structure of the particular subject still awaits development and clarification.

Many mathematicians have contributed to the general results now available in the theory of deficiency indices of ordinary differential operators. Some mention is made here of certain individual contributions in order to place them in historical perspective; for a more detailed historical account, but only up to about 1955, see [9, Chapter XIII, Section 10A].

Following the original work of Weyl [57] in 1910 the next outstanding contributions came from Stone [52] in 1932, and Titchmarsh from 1939 onwards whose results are collected in his book [55]; in particular Titchmarsh's contribution is important for his penetrating use of the theory of (Cauchy) analytic functions in his analysis of the limit-point and limit-circle cases. Kodaira [35] in 1949, Glazman [22] in 1950 and Naimark [42] in 1952 all made significant advances; in particular Glazman solved completely the range problem for deficiency indices of real, even-order differential expressions. The bounds (2.13), (2.14) and (2.15) for the deficiency indices in the general order complex case seem first to have been obtained by Everitt [14] in 1959. Kimura and Takahasi [32] in 1965 extended the Kodaira analysis to the general complex case (it should be noted that the second part of this work has not appeared). In 1966 McLeod [41a] gave the first example of a complex fourth-order differential expression with unequal deficiency indices; this was followed in 1971 by the contribution from Kogan and Rofe-Beketov [36] in which the partial solution to the range problem was given, i.e. all cases are realizable subject to the additional constraint $|N_+ - N_-| \leq 1$. The very recent work of Gilbert [21b] goes a long way to remove this constraint which is now seemingly unnecessary.

The coefficient problem in the theory of deficiency indices of ordinary differential operators continues in a steady state of advancement. This concerns the search for necessary and sufficient conditions on the coefficients of the differential expression $w^{-1}M$ on I to determine the deficiency indices. This problem is implicit in the 1910 paper [57] of Hermann Weyl; see the remarks made in Section 1 above. For general real even-order differential expressions the problem was first seriously considered by Glazman [22] and Naimark [42] from 1950 onwards. In the general complex case the problem was stated by Coddington [5] in 1954.

We mention here only a small number of important contributions to the coefficient problem for higher even-order symmetric differential expressions. Ismagilov [27] generalized earlier results to higher order expressions; Devinatz [7] and Eastham [11], [12] have obtained many significant results in the fourth-order real case. Kogan and Rofe-Beketov [36], [37] and Rofe-Beketov [49] have done much to advance knowledge of the general complex case. An outstanding contribution was made by Atkinson [2] in the study of general differential expressions.

An interesting conjecture in the so-called 'positive coefficient' case was shown to be false by Kauffman [31] in 1976. If $w = 1$ and M, of even-order 2r, is given on $[0,\infty)$ in the form

$$M[f] = \sum_{s=0}^{r} (-1)^s (p_s f^{(s)})^{(s)}$$

with $p_r > 0$ and $p_s \geq 0$ $(s = 0,1,\ldots,r-1)$ it was conjectured by Everitt in 1968 that the deficiency indices of M in $L^2(0,\infty)$ (necessarily equal since M is real) should be in the minimal position, i.e. $N_+ = N_- = r$. Kauffman [31] has now shown that this conjecture is false when $r = 3$, i.e. $n = 6$, and it seems likely that the conjecture is false in general; however the fourth-order case is still not settled. J. B. McLeod has advanced strong reasons for supposing that in the positive coefficient case all values of $N_+ = N_-$ are possible except the maximal case, i.e. $r \leq N_+ = N_- \leq 2r - 1 < 2r$ represents the complete state of affairs for all $n = 2r \geq 2$.

To conclude this section we mention two interesting but very different proble

(i) can the fourth-order real symmetric differential expression $f^{(4)} + qf$ on $[0,\infty)$ be in the maximal case, i.e. $N_+ = N_- = 4$ (it is known that, by choice of the coefficient q, the cases $N_+ = N_- = 2$ or 3 are both possible)?

(ii) can the operator theoretic methods of Glazman [22], by means of which h solved the range problem in the real even-order case, be extended to solve partial or completely the range problem in the general complex case?

Section 3

In this last section we mention briefly certain ramifications of the deficien index problem

Left-definite problems In the second-order case these are concerned with the differential equation

$$-(py')' + qy = \lambda wy \text{ on } I$$

when w is no longer of one sign on I, but both p and q are non-negative on I. Problems of this kind are originally considered by Weyl, see the note at the end of [57], and called the polar case on the basis of an earlier terminology of Hilbert A framework for these problems is the Hilbert function space with inner-product

$$(f,g) = \int_I \{pf'\overline{g}' + qf\overline{g}\}.$$

In this space the differential equation may be classified as limit-point or limit-circle although the situation is somewhat more complicated than in the right-definite case when w is non-negative on I. For some discussion of the Titchmarsh-Weyl theory of these problems see Atkinson, Everitt, Ong [3] which has references to earlier work. For an interesting and important result on the limit-circle case see Niessen [43] which also contains information concerning other results in this field.

Pairs of symmetric differential expressions. The left-definite problem mentioned above forms a special case of the more general differential equation

$$Sy = \lambda Ty \quad \text{on} \quad I$$

where both S and T are symmetric differential expressions with the order of T less than the order of S. In recent years Pleijel and his school at the University of Uppsala have made distinguished contributions to the study of this equation and the resulting spectral theory. Either S or T is chosen to be definite and the resulting cases are called left- or right-definite respectively. Reference should be made to the survey works of Pleijel [45], [46]. An interesting proof of the bounds and related results for the deficiency indices of this problem, equivalent to the results given in (2.12 to 15), is given by Bennewitz in [4]. There is a similar extension of the result (2.11) to be found in the work of Karlsson [29].

It is much to be hoped that a collected account of the work of Pleijel and the Uppsala school will be written in due course, in order that these important results may be found together in one reference source.

Strong limit-point and Dirichlet results. The original limit-point case of the second-order differential expression considered by Weyl [57] has been further classified in recent years to the strong and weak limit-point cases. The strong limit-point case is intimately connected with the so-called Dirichlet condition of the differential expression concerned. For an account of the definitions involved and a survey of results in this area see the paper by Everitt [18]

Powers of symmetric differential expressions. If the differential expression M of order n is symmetric then it is possible to consider powers M^r of M for positive integers r. Formally these are defined by

$$M^1 = M, \quad M^2 = M[M], \quad \ldots, \quad M^r = M[M^{r-1}]$$

but such a definition requires heavy differentiability conditions on the coefficients of M. Recently Zettl [62] has shown that it is possible to form all powers M^r of quasi-differential expressions with no additional restrictions on the coefficients. It then is the case that M^r is also a symmetric quasi-differential expression of order rn.

In recent years properties of the powers M^r ($r = 2,3,\ldots$) have been extensively studied and much is now known of the relationships between the deficiency indices of M and those of M^r. For a survey of results up to 1974 see the paper by Everitt and Giertz (with addenda by Kauffman, Read and Zettl) [19].

Special mention should be made of the work of Kauffman [30] and Read [48] in which the range problem of deficiency indices of powers of real symmetric differential expressions was solved.

A collected account of many of these results has now been prepared by Kauffman, Read and Zettl and this should be published shortly under the title "The deficiency index problem for powers of ordinary differential expressions".

Conclusion. The deficiency index problem for ordinary differential operators is an outstanding example in which ideas old and new, methods classical and modern interplay one with the other to the permanent advantage of the development of mathematica analysis.

Towards the end of his long life, and as mentioned in Section 1, Hermann Weyl looked back over the years in an address to the American Mathematical Society [58] entitled "Ramifications, old and new, of the eigenvalue problem". On commenting on the importance of the interplay between concrete and abstract problems in mathematics he wrote of the development of the singular differential equation problem during the years from 1910 to 1950 (see [58, Page 441])

"It is remarkable that forty years had to pass before such a thoroughly satisfactory direct treatment emerged; the fact is a reflection on the degree to which mathematicians during this period got absorbed in abstract generalizations and lost sight of their task of finishing up some of the more concrete problems of undeniable importance."

References

1. F. V. Atkinson. Discrete and continuous boundary problems. (Academic Press, New York, 1964).

2. F. V. Atkinson. 'Limit-n criteria of integral type'. Proc. Royal Soc. Edinburgh (A) 73 (1975), 167-198.

3. F. V. Atkinson, W. N. Everitt and K. S. Ong. 'On the m-coefficient of Weyl for a differential equation with an indefinite weight function'. Proc. London Math. Soc. (3) 29 (1974), 368-384.

4. C. Bennewitz. 'Generalisation of some statements of Kodaira'. Proc. Royal Soc. Edinburgh (A) 71 (1972/73), 113-119.

5. E. A Coddington. 'The spectral representation of ordinary self-adjoint differential operators'. Ann. of Math. 60 (1954), 192-211.

6. E. A. Coddington. 'Spectral theory of ordinary differential operators'. Lecture Notes in Mathematics (Springer Verlag, Berlin) 448 (1975), 1-24.

7. A. Devinatz. 'On a deficiency index theory of W. N. Everitt'. Lecture Notes in Mathematics (Springer-Verlag, Berlin) 415 (1974), 103-108.

8. A. Devinatz. 'The deficiency index problem for ordinary self-adjoint differential operators'. Bull. Amer. Math. Soc. 79 (1973), 1109-1127.

9. N. Dunford and J.T. Schwartz. Linear operators : II (Interscience, New York, 1963).

10. M. S. P. Eastham and M. L. Thompson. 'On the limit-point, limit-circle classification of second-order ordinary differential equations'. Quart. J. Math. (Oxford) (2) 24 (1973), 531-35.

11. M. S. P. Eastham. 'The limit-3 case of self-adjoint differential expressions of the fourth-order with oscillating coefficients'. J. London Math. Soc.(2) 8 (1974), 427-37.

12. M. S. P. Eastham. 'Square-integrable solutions of the differential equation $y^{(4)} + a(qy')' + (bq^2 + q'')y = 0$'. To appear in Niew Archief voor Wiskunde.

13. W. D. Evans. 'On limit-point and Dirichlet-type results for second-order differential expressions. Lecture Notes in Mathematics (Springer-Verlag, Berlin) 564 (1976), 78-92.

14. W. N. Everitt. "Integrable-square solutions of ordinary differential equations'. Quart. J. Math. (Oxford) (2) 10 (1959), 145-155.

15. W. N. Everitt. 'A note on the self-adjoint domains of second-order differential equations' Quart. J. Math. (Oxford) (2) 14 (1963), 41-45.

16. W. N. Everitt. 'Integrable-square solutions of ordinary differential equations (III)'. Quart. J. Math. (Oxford) (2) 14 (1963), 170-180.

17. W. N. Everitt. 'Integrable-square, analytic solutions of odd-order, formally symmetric, ordinary differential equations'. Proc. London Math. Soc. (3) 25 (1972), 156-182.

18. W. N. Everitt. 'On the strong limit-point condition of second-order differenti expressions'. Proceedings of the International Conference on Differential Equations, Los Angeles. (1974), 287-307. (Academic Press, Inc., New York).

19. W. N. Everitt and M. Giertz. 'On the deficiency indices of powers of formally symmetric differential expressions'. Lecture Notes in Mathematics (Springer-Verlag, Berlin) 444 (1974), 167-181. (With addenda by R. Kauffman, T. T. Read and A Zettl.)

20. W. N. Everitt and V. Krishna Kumar. 'On the Titchmarsh-Weyl theory of ordinary symmetric differential expressions I : the general theory'. Nieuw Archief voor Wiskunde (3) XXXIV (1976), 1-48.

21. W. N. Everitt and V. Krishna Kumar. 'On the Titchmarsh-Weyl theory of ordinary symmetric differential expressions II : the odd-order case'. Nieuw Archief voor Wiskunde (3) XXXIV (1976), 109-145.

21a. M. Giertz, 'Report from the conference on ordinary and partial differential equations held in Dundee, March 30 to April 2, 1976' TRITA-MAT-1976-7 Mathematics (Royal Institute of Technology, Stockholm, Sweden).

21b. R. C. Gilbert. 'On deficiency indices of ordinary differential operators'. J. Diff. Equations (to appear).

22. I. M. Glazman. 'On the theory of singular differential operators' Uspehi Math. Nauk 40 (1950), 102-135. (English translation in Amer. Math. Soc. Transl.(1) 4 (1962), 331-372.

23. I. Halperin. 'Closures and adjoints of linear differential operators'. Ann. of Math. 38 (1937), 880-919.

24. P. Hartman. 'The number of L^2-solutions of $x'' + q(t)x = 0$'. Amer. J. of Math. 73 (1951), 635-645.

25. P. Hartman and A. Winter. 'A criterion for the non-degeneracy of the wave equation'. Amer. J. Math. 71 (1949), 206-213.

26. E. Hille. Lectures on ordinary differential equations (Addison-Wesley, London, 1969).

27. R. S. Ismagilov. 'Conditions for self-adjointness of differential equations of higher order'. Soviet Math. 3 (1962), 279-283.

28. R. S. Ismagilov. 'On the self-adjointness of the Sturm-Liouville operator'. Uspehi Mat. Nauk. 18 (1963), 161-166.

29. B. Karlsson. 'Generalization of a theorem of Everitt'. J. London Math. Soc. (2) 9 (1974), 131-141.

30. R. M. Kauffman. 'A rule relating the deficiency indices of L^j to those of L^k'. Proc. Royal Soc. Edinburgh (A) 74 (1976), 115-118.

31. R. M. Kauffman. 'On the limit-n classification of ordinary differential operators with positive coefficients'. Lecture Notes in Mathematics (Springer-Verlag, Berlin) 564 (1976), 259-266.

32. T. Kimura and M. Takahasi. 'Sur les opérateurs différentiels ordinaires linéaires formellement autoadjoints I'. Funkcialoj Ekvacioj (Ser. Int.) 7 (1965), 35-90.

33. I. W. Knowles. 'Note on a limit-point criterion'. Proc. Amer. Math. Soc. 41 (1973), 117-119.

34. I. W. Knowles. The limit-point and limit-circle classification of the Sturm-Liouville operator (py')' + qy. (Ph.D. thesis, Flinders University of South Australia : 1972.)

35. K. Kodaira. 'On ordinary differential equations of any even order and the corresponding eigenfunction expansion'. Amer. J. Math. 71 (1949), 921-45.

36. V. I. Kogan and F.S.Rofe-Beketov. 'On the question of the deficiency indices of differential operators with complex coefficients'. Mat. Fiz. Funk. Anal. 2 (1971), 45-60. (Russian; Akad. Nauk. Ukr. SSR, Fiz-Tek. Inst. Nizk. Temp., Kharkov : English translation; Proc. Royal Soc. Edinburgh (A) 72 (1973/74), 281-198.)

37. V. I. Kogan and F. S. Rofe-Beketov. 'On square-integrable solutions of symmetric systems of differential equations of arbitrary order'. Proc. Royal Soc. Edinburgh (A) 74 (1974/75), 5-40.

38. M. K. Kwong. 'L^p-perturbations of second-order linear differential equations'. Math. Ann. 215 (1975), 23-34.

39. N. Levinson. 'Criteria for the limit-point case for second-order linear differential operators'. Casopis pro pesto vanyi matematiky a fysiky'. 74 (1949), 17-20.

40. B. M. Levitan and I. S. Sargsjan. Self-adjoint ordinary differential operators (English translation from the Russian in Translations of Mathematical Monographs, 39, Amer. Math. Soc. 1975.)

41. J. B. McLeod. 'Square-integrable solutions of a second-order differential equation with complex coefficients'. Quart. J. Math. (Oxford) 13 (1962), 129-133.

41a. J. B. McLeod. 'The number of integrable-square solutions of ordinary differential equations'. Quart. J. Math. (Oxford) 17 (1966), 285-290.

42. M. A. Naimark. Linear differential operators (English translation from the Russian : Ungar, New York; Part I, 1967; Part II, 1968.)

43. H.-D. Niessen. 'A necessary and sufficient limit-circle criterion for left-definite eigenvalue problems'. Lecture Notes in Mathematics (Springer-Verlag, Berlin) 415 (1974), 205-210.

44. J. Nitsche. 'Über systeme kanonischer Differentialgleichungen und das zugehörge singuläre eigenwertproblem'. Wissenschaftliche Zeitschrift der UniversitätLeipzig 5 (1952/53), 193-226.

45. Åke Pleijel. 'Generalized Weyl circles'. Lecture Notes in Mathematics (Springer-Verlag, Berlin) 415 (1974), 211-226.

46. Åke Pleijel. 'A survey of spectral theory for pairs of ordinary differential operators'. Lecture Notes in Mathematics (Springer-Verlag, Berlin) 448 (1975), 256-272.

47. D. Race. Limit-point and limit-circle : 1910-1970 (MSc thesis, University of Dundee, Scotland, UK : 1976).

48. T. T. Read. 'Sequences of deficiency indices'. Proc. Royal Soc. Edinburgh (A) 74 (1974/75), 157-164.

49. F. S. Rofe-Beketov. 'Deficiency indices and properties of the spectrum of some classes of differential operators'. Lecture Notes in Mathematics (Springer-Verlag, Berlin) 448 (1975), 273-293.

50. D. B. Sears. 'Note on the uniqueness of Green's functions associated with certain differential equations'. Canadian J. Math. 2 (1950), 314-325.

51. D. B. Sears. 'Some properties of a differential equation'. J. London Math. Soc. 27 (1952), 180-188.

52. D. Shin. 'On quasi-differential operators in Hilbert space'. Doklad. Akad. Nauk SSSR 18 (1938), 523-6. (Russian.)

53. D. Shin. 'On the solutions of a linear quasi-differential equation of the n'th order' Mat. Sb. 7 (1940), 479-532. (Russian.)

54. M. H. Stone. Linear transformations in Hilbert space and their applications to analysis. (Amer. Math. Soc. Coll. Publ. 15, New York, 1932.)

55. E. C. Titchmarsh. Eigenfunction expansions : I (Oxford University Press, 1962).

56. P. W. Walker. 'A vector-matrix formulation for formally symmetric ordinary differential equations with applications to solutions of integrable square'. J. London Math. Soc. (2) 9 (1974), 151-159.

57. H. Weyl. 'Über gewöhnliche Differentialgleichungen mit Singularitäten und die zugehörigen Entwicklungen willkürlicher Funktionen. Math. Ann. 68 (1910), 220-269.

58. H. Weyl. 'Ramifications, old and new, of the eigenvalue problem'. Bull. Amer. Math. Soc. 56 (1950), 115-139.

59. J. S. W. Wong. 'Square integrable solutions of L^p perturbations of second-order linear differential equations'. Lecture Notes in Mathematics (Springer-Verlag, Berlin) 415 (1974), 282-292.

60. K. Yosida. Lectures on differential and integral equations (Interscience, New York, 1960).

61. A. Zettl. 'Formally self-adjoint quasi-differential operators'. Rocky Mountain J. Math. 5 (1975), 453-474.

62. A Zettl. 'Powers of real symmetric differential expressions without smoothness assumptions'. Quaestiones Mathematicae 1 (1976), 83-94.

SINGULAR PERTURBATIONS OF THE PRINCIPAL EIGENVALUE OF ELLIPTIC OPERATORS

Avner Friedman

Northwestern University, Evanston, Illinois

1. Definition of the principal eigenvalue. Let

$$(1.1)\qquad Lu \equiv \frac{1}{2}\sum_{i,j=1}^{n} a_{ij}(x)\frac{\partial^2 u}{\partial x_i \partial x_j} + \sum_{i=1}^{n} b_i(x)\frac{\partial u}{\partial x_i}$$

be an elliptic operator with C^1 coefficients in R^n and let Ω be a bounded domain with C^2 boundary $\partial\Omega$. Consider the eigenvalue problem

$$(1.2)\qquad \begin{aligned} -Lu &= \lambda u \quad \text{in } \Omega,\\ u &= 0 \quad \text{on } \partial\Omega. \end{aligned}$$

From the theory of positive operators [10] it is known that there exists a positive eigenvalue λ_0 with eigenfunction $\phi_0(x)$ which is positive throughout Ω, and (i) $|\lambda| > \lambda_0$ for any eigenvalue λ, (ii) λ_0 is a simple eigenvalue. It is further known [12] that $\text{Re}\,\lambda \geq \lambda_0$ for any other eigenvalue. The number λ_0 is called the principal eigenvalue.

2. Properties of λ_0. Let $(\sigma_{ij}(x))$ be an $n\times n$ matrix such that $\sigma\sigma^* = a$ (σ^* = transpose of σ) where $a = (a_{ij})$, and set $b = (b_1,\ldots,b_n)$. Consider the system of stochastic differential equations (for definition see, for instance, [6])

$$(2.1)\qquad d\xi(t) = \sigma(\xi(t))dw(t) + b(\xi(t))dt$$

where $w(t)$ is an n-dimensional Brownian motion. Denote by τ_x the exit time of $\xi(t)$ from Ω, given that $\xi(0) = x$. Set

$$(2.2)\qquad \Lambda = \sup\{\lambda \geq 0;\ \sup_{x\in\Omega} Ee^{\lambda\tau_x} < \infty\}.$$

Theorem 2.1. $\Lambda = \lambda_0$.

This probabilistic characterization is due to Khasminskii [8]; for proof see also [7].

The following comparison theorem is an immediate consequence of Theorem 2.1.

Theorem 2.2. *If $\Omega' \subset \Omega$ and λ_0' is the principal eigenvalue of L corresponding to Ω' (Ω' is a domain with C^2 boundary), then $\lambda_0 \leq \lambda_0'$.*

A useful tool in estimating λ_0 is given by:

Theorem 2.3. (i) *Suppose there exists a function $\Psi(x)$ in $C(\bar\Omega) \cap C^2(\Omega)$ such that*

This work was partially supported by National Science Foundation Grant MCS75-21416 A01.

$$L\Psi + A\Psi \geq 0 \text{ in } \Omega, \quad A > 0,$$

$$\Psi(x) \geq 0 \text{ in } \Omega, \quad \Psi(x) = 0 \text{ on } \partial\Omega.$$

Then $\lambda_0 \leq A$.

(ii) Suppose there exists a function $\Phi(x)$ in $C(\bar{\Omega}) \cap C^2(\Omega)$ such that

$$L\Phi + B\Phi \leq 0 \text{ in } \Omega, \quad B > 0,$$

$$\Phi(x) \geq 1 \text{ in } \Omega.$$

Then $\lambda_0 \geq B$.

This theorem is introduced in [1].

Suppose we have a family of positive definite matrices $(a_{ij}(x,\varepsilon))$ in $\bar{\Omega}$ ($\varepsilon > 0$) such that $a_{ij}(x,\varepsilon) \to a_{ij}(x,0)$ as $\varepsilon \to 0$, uniformly in $\bar{\Omega}$. Denote by λ_ε the principal eigenvalue of

$$M_\varepsilon u \equiv \frac{1}{2} \sum_{i,j=1}^{n} a_{ij}(x,\varepsilon) \frac{\partial^2 u}{\partial x_i \partial x_j} + \sum_{i=1}^{n} b_i(x) \frac{\partial u}{\partial x_i} . \tag{2.3}$$

If $(a_{ij}(x,0))$ is non-degenerate then we have (see [1]):

$$\lambda_\varepsilon \to \lambda_0 \quad \text{if } \varepsilon \to 0, \tag{2.4}$$

where λ_0 is the principal eigenvalue of M_0. In what follows we shall be interested in the case where M_0 is a degenerate elliptic operator so that a principal eigenvalue λ_0 does not exist (for instance, if $a_{ij}(x,0) \equiv 0$). We shall study the question:

How does λ_ε behave as $\varepsilon \to 0$?

3. The case $b\cdot\nu < 0$. Denote by ν the outward normal to $\partial\Omega$, and assume that

$$b\cdot\nu < 0 \quad \text{along } \partial\Omega. \tag{3.1}$$

Denote by λ_ε the principal eigenvalue of

$$\begin{aligned} L_\varepsilon u \equiv \frac{\varepsilon^2}{2} \sum_{i,j=1}^{n} a_{ij}(x) \frac{\partial^2 u}{\partial x_i \partial x_j} + \sum_{i=1}^{n} b_i(x) \frac{\partial u}{\partial x_i} &= -\lambda u \quad \text{in } \Omega, \\ u = 0 \quad \text{on } \partial\Omega. \end{aligned} \tag{3.2}$$

We assume that (a_{ij}) is positive definite in $\bar{\Omega}$ and that a_{ij}, b_i are continuously differentiable in R^n. The condition (3.1) implies that the solution of

$$\frac{dx}{dt} = b(x) \quad \text{with } x(0) \in \Omega \tag{3.3}$$

does not leave Ω at any time. Consequently, for the solution of the stochastic system

$$d\xi(t) = \varepsilon\sigma(\xi(t))dw(t) + b(\xi(t))dt, \tag{3.4}$$

the exit time τ_x^ε increases to ∞ as $\varepsilon \to 0$. Theorem 2.1 then implies that (at least heuristically) λ_ε should go to zero as $\varepsilon \to 0$. Indeed, one can show (see [4]) that, for small ε,

$$\lambda_\varepsilon \leq e^{-c/\varepsilon^2} \quad \text{for some } c > 0. \tag{3.5}$$

A more precise assertion can be stated under some additional assumptions, which we proceed to describe.

Denote by $(\tilde{a}_{ij})$ the inverse matrix to (a_{ij}), and let

$$I_T(\phi) = \int_0^T \sum_{i,j=1}^n \tilde{a}_{ij}(\phi(t))\left(\frac{d\phi^i}{dt} - b^i(\phi(t))\right)\left(\frac{d\phi^j}{dt} - b^j(\phi(t))\right)dt$$

where $\phi = (\phi^1,\dots,\phi^n)$ is any absolutely continuous curve. For any two points x,y in Ω, let

$$V(x,y) = \inf I_T(\phi),$$

where the "inf" is taken over all absolutely continuous curves $\phi(t)$ such that $\phi(0) = x$, $\phi(T) = y$ and $\phi(t) \in \Omega$ for all $0 < t < T$. We now assume:

(i) There exists a finite number of disjoint compact sets $K_1,\dots,K_\ell$ in Ω such that the ω-limit set of each solution of (3.3) with $x(0) \in \Omega \setminus (\bigcup_{i=1}^{\ell} K_i)$ is contained in one of the sets K_i.

(ii) $V(x,y) = 0$ if x and y belong to the same set K_j.

Set

$$V_i = \inf_{y \in \partial\Omega} V(x,y), \quad x \in K_i,$$

$$V^* = \max(V_1,\dots,V_\ell), \quad V_* = \min(V_1,\dots,V_\ell).$$

Then we have:

Theorem 3.1. *If (i),(ii) hold, then*

$$\overline{\lim_{\varepsilon\to 0}}(-2\varepsilon^2 \log \lambda_\varepsilon) \le V^*, \tag{3.6}$$

$$\underline{\lim_{\varepsilon\to 0}}(-2\varepsilon^2 \log \lambda_\varepsilon) \ge V_*. \tag{3.7}$$

For proof see Friedman [4][7] and Ventcel [13][14].

The proof exploits Theorem 1.1 and probabilistic estimates of Ventcel and Freidlin [16].

4. The case when all solutions of (3.3) leave $\bar{\Omega}$. We now replace (3.1) by the other extreme assumption that all solutions of (3.3) with $x(0) \in \Omega$ leave $\bar{\Omega}$ in finite time. Theorem 2.1 then indicates that λ_ε should tend to ∞ as $\varepsilon \to 0$. Indeed, we have:

Theorem 4.1. *If all solutions of (3.3) leave $\bar{\Omega}$ in finite time, then*

$$2\varepsilon^2\lambda_\varepsilon \to c \quad \text{as } \varepsilon \to 0 \tag{4.1}$$

where c is a positive constant given by

$$c = \lim_{T\to\infty} \frac{1}{T} \inf\{I_T(\phi);\ \phi(t) \in \bar{\Omega} \text{ for all } 0 \le t \le T\}. \tag{4.2}$$

This theorem, which is due to Ventcel [15], also exploits the Ventcel-Freidlin probabilistic estimates.

5. Cases when $b\cdot\nu > 0$. In this section we consider an intermediate case, namely, $b\cdot\nu > 0$ but not all solutions of (3.3) leave $\bar{\Omega}$. In fact, we specialize to the case where there is a point x^0 in Ω such that

(5.1) $\quad b(x)$ has a zero of order μ at x^0; $\mu \geq 1$.

This ensures that $x(t) \equiv x^0$ is a solution of (3.3) (which, of course, does not leave Ω).

Theorem 5.1. If (5.1) holds, then

$$\lambda_\varepsilon \leq C\varepsilon^{2(\mu-1)/(\mu+1)} \qquad (C > 0) \tag{5.2}$$

if ε is sufficiently small.

Assume next that x^0 is "globally repulsive," i.e.,

(5.3)
there exists a function $\psi \in C^2(\bar{\Omega})$ such that
$b\cdot\nabla\psi \geq k|x - x^0|^{1+\mu}$ in $\bar{\Omega}$, $k > 0$,
$(\psi_{x_i x_j}(x^0))$ is positive definite.

Theorem 5.2. If (5.3) holds, then

$$\lambda_\varepsilon \geq c\varepsilon^{2(\mu-1)/(\mu+1)} \qquad (c > 0) \tag{5.4}$$

if ε is sufficiently small.

Theorems 5.1, 5.2 are given in [1]; they are valid also when the b_i are just Hölder continuous and when μ is any non-negative number. In the special case $\mu = 0$, the result is not as sharp as that in Theorem 4.1.

The proof of Theorems 5.1, 5.2 exploits Theorem 2.3. Another proof (also given in [1]) is based on generalizing the approach of the variational principle associated with the first eigenvalue in the selfadjoint case (to the present not necessarily selfadjoint case).

6. The limit operator is degenerate elliptic. We now consider the eigenvalue problem

$$\begin{aligned} \varepsilon\Delta u + Lu &= -\lambda u \quad \text{in } \Omega, \\ u &= 0 \quad \text{on } \partial\Omega \end{aligned} \tag{6.1}$$

where L is the elliptic operator in (1.1) with continuously differentiable coefficients in $\bar{\Omega}$; we assume that L does not degenerate in Ω but that it degenerates on $\partial\Omega$. [All the considerations below extend to the case where Δ is replaced by any other non-degenerate elliptic operator.] What we have in mind is a degeneracy of the type:

$$\sum_{i,j=1}^{n} a_{ij}\nu_i\nu_j = 0 \quad \text{on } \partial\Omega, \tag{6.2}$$

(6.3) $$\sum_{i=1}^{n} \Big(b_i - \frac{1}{2} \sum_{j=1}^{n} \frac{\partial a_{ij}}{\partial x_j}\Big)\nu_i \leq 0 \quad \text{on } \partial\Omega$$

where $\nu = (\nu_1,\ldots,\nu_n)$ is the outward normal. This means (see [7]) that the exit time τ_x for the solution $\xi(t)$ of (2.1) is $= \infty$. For the stochastic solution $\xi_\varepsilon(t)$ corresponding to $\varepsilon\Delta + L$ it means that the exit time $\tau_x^\varepsilon \to \infty$ if $\xi(t)$ stays "well within" Ω for a "long" time. This intuition can be put in quantitative terms. Results to this effect are given in [3] and involve the quantities

(6.4) $$\mathcal{A}(x) = \frac{1}{2} \sum_{i,j=1}^{n} a_{ij}(x) \frac{\partial\rho(x)}{\partial x_i} \frac{\partial\rho(x)}{\partial x_j},$$

(6.5) $$\mathcal{B}(x) = \sum_{i=1}^{n} \Big(b_i(x) - \frac{1}{2} \sum_{j=1}^{n} \frac{\partial a_{ij}(x)}{\partial x_j}\Big)\frac{\partial\rho(x)}{\partial x_i}$$

where $\rho(x)$ = distance from x to $\partial\Omega$.

For upper bounds on λ_ε we assume:

(6.6) $$\mathcal{A}(x) \leq a\rho^k, \quad a > 0,\ k > 0,$$

(6.7) $$\mathcal{B}(x) \geq \hat{b}\rho^\beta, \quad \hat{b} > b > 0,\ \beta > 0$$

in some Ω-neighborhood of $\partial\Omega$, and set

$$I_{\beta,k} = \frac{k^\beta}{\Gamma(\frac{1}{k})} \Gamma(1 + \beta + \frac{1}{k}) \frac{\pi}{\sin \frac{\pi(1+\beta)}{k}},$$

$$c = bI_{\beta,k}/a^{(\beta+1)k}, \quad \mu = 1 - \frac{\beta+1}{k}, \quad \nu = (\frac{b}{a} - 1)/k \quad \text{if } k \geq 1+\beta.$$

<u>Theorem 6.1</u>. <u>If</u> (6.6),(6.7) <u>hold then, for some positive constant</u> A,

(6.8) $$\lambda_\varepsilon \leq Ae^{-c/\varepsilon^\mu} \quad \underline{\text{if}}\ k > 1+\beta,$$

(6.9) $$\lambda_\varepsilon \leq \left\{ \begin{array}{ll} A\varepsilon^\nu & \underline{\text{if}}\ \frac{b}{a} > 1 \\ A & \underline{\text{if}}\ \frac{b}{a} < 1 \end{array} \right\} \quad \underline{\text{for}}\ k = 1+\beta$$

<u>for all</u> ε <u>sufficiently small</u>.

For lower bounds on λ_ε we assume that there exists a C^2 function $p(x)$ in $\bar{\Omega}$ which coincides with dist.$(x,\partial\Omega)$ in some Ω-neighborhood of $\partial\Omega$, such that $p(x)$ has just one critical point x^0 in Ω and the Hessian is negative definite at x^0. (Such a function always exists if Ω is diffeomorphic to a ball.)

We also assume that in some Ω-neighborhood of $\partial\Omega$,

(6.10) $$\mathcal{A}(x) \geq a\rho^k, \quad a > 0,\ k > 0,$$

(6.11) $$\mathcal{B}(x) \leq \tilde{b}\rho^\beta, \quad 0 < \tilde{b} < b,\ \beta \geq 0,$$

and

(6.12) $$\sum a_{ij} \frac{\partial^2 \rho}{\partial x_i \partial x_j} = -\sum \frac{\partial a_{ij}}{\partial x_j} \frac{\partial \rho}{\partial x_i} + o(\rho^\beta) \quad \text{if } \beta > 0,\ \rho \to 0,$$

(6.13) $$\sum a_{ij} \frac{\partial \rho}{\partial x_i} \frac{\partial \rho}{\partial x_j} = 0 \quad \text{on } \partial\Omega \text{ if } \beta = 0.$$

The condition (6.13) coincides with (6.2) and the condition (6.12) with $\beta < 1$ follows from (6.2) if $a_{ij} \in C^2$, $\rho \in C^3$ in some $\bar{\Omega}$-neighborhood of $\partial\Omega$ (see [6, p. 208]).

Theorem 6.2. Under the assumptions (6.10)-(6.13), there exists a positive constant A such that

(6.14) $$\lambda_\varepsilon \geq Ae^{-c/\varepsilon^\mu} \quad \text{if } k > 1+\beta$$

(6.15) $$\lambda_\varepsilon \geq \begin{cases} A\varepsilon^\nu & \text{if } \frac{b}{a} > 1 \\ A & \text{if } \frac{b}{a} < 1 \end{cases} \quad \text{for } k = 1+\beta,$$

(6.16) $$\lambda_\varepsilon \geq A \quad \text{if } k < 1+\beta$$

for all ε sufficiently small.

The proofs of Theorems 6.1, 6.2 (in [3]) make use of Theorem 2.3 and involve the construction of appropriate comparison functions. The estimates of Theorems 6.1 and 6.2 complement each other and are rather sharp.

7. Principal eigenvalues for degenerate operators.

In various problems which arise in genetics and in population dynamics, one encounters the eigenvalue problem (3.2) with L which degenerates on the boundary. A natural question to ask is then whether a principal eigenvalue exists in some reasonable sense. This is considered in [2] in the case of ordinary differential equations. (The results of [2] extend to partial differential equations of a very restricted form.) The eigenvalue problem is then

(7.1) $$L_\varepsilon u \equiv \frac{\varepsilon^2}{2} a(x)u'' + b(x)u' = -\lambda u \qquad (0 < x < 1),$$
$$u(0) = u(1) = 0$$

where a, b are continuously differentiable functions in (0,1), $a(x) > 0$, and it is assumed that

(7.2) $$\frac{a(x)}{x^\alpha} \to a_0 > 0, \quad \frac{b(x)}{x^\beta} \to b_0 \quad \text{as } x \to 0,$$
$$\frac{a(x)}{(1-x)^{\alpha'}} \to a_1 > 0, \quad \frac{b(x)}{(1-x)^{\beta'}} \to b_1 \quad \text{as } x \to 1,$$
$$0 \leq \alpha,\ \alpha' < 2; \quad \beta, \beta' > 0, \quad \alpha < 1+\beta, \quad \alpha' < 1+\beta'.$$

Let

$$p_\varepsilon(x) = \exp\left[\int_0^x \frac{2b(t)}{\varepsilon^2 a(t)}\, dt\right].$$

Theorem 7.1. There exists an eigenvalue λ_ε and eigenfunction u_ε of (7.1) such that $u_\varepsilon(x) > 0$ if $0 < x < 1$, and u_ε is Hölder continuous in $[0,1]$ with exponent 1/2. Fo any solution v of

$$L_\varepsilon v = -\mu v \quad \text{in } (0,1), \quad \mu \text{ real},$$

$$v(0) = v(1) = 0 \quad (v \text{ continuous in } [0,1]),$$

$$\frac{p_\varepsilon v^2}{a} \in L^2(0,1)$$

we have $\mu \geq \lambda_\varepsilon$. Finally, if $a(x)/x^\alpha$ is in C^2 in some interval $0 \leq x < \delta$, then the eigenvalue λ_ε is simple.

We call λ_ε the principal eigenvalue.

We next assume that

$$\text{(7.3)} \quad \begin{aligned} &b_0 > 0,\ b_1 < 0; \text{ there exists an } \hat{x} \in (0,1) \text{ such that} \\ &b(x) > 0 \text{ if } 0 < x < \hat{x}, \quad b(x) < 0 \quad \text{if } \hat{x} < x < 1. \end{aligned}$$

We set

$$\text{(7.4)} \qquad J = \min\left\{\int_0^{\hat{x}} \frac{4b(t)}{a(t)}\,dt,\ \int_1^{\hat{x}} \frac{4b(t)}{a(t)}\,dt\right\}.$$

Theorem 7.2. If (7.3) holds, then

$$\text{(7.5)} \qquad -2\varepsilon^2 \log \lambda_\varepsilon \to J \quad \text{as } \varepsilon \to 0.$$

The proof is based on considering the eigenvalue problem for $\eta u'' + L_\varepsilon u$ $(\eta > 0)$ and studying carefully the behavior of its principal eigenvalue as $\eta \to 0$, $\varepsilon \to 0$ (exploiting Theorem 3.1).

For details regarding the proofs of Theorems 7.1, 7.2 see [2]. Analogs of Theorems 5.1, 5.2 are also valid for the present L_ε, but the proofs are straightforward generalizations of the proofs of Theorems 5.1, 5.2.

8. Open questions.

1. Study the behavior of the principal eigenfunction $\phi_\varepsilon(x)$ as $\varepsilon \to 0$. This is related to the asymptotic behavior of Green's function $q_\varepsilon(x,y,t)$ of the correspondir parabolic equation in the cylinder $\Omega \times (0,\infty)$. In [5] it was shown that

$$\text{(8.1)} \qquad -2\varepsilon^2 q_\varepsilon(x,y,t) \to I_t(x,y) \quad \text{as } \varepsilon \to 0,$$

where $I_t(x,y) = \inf I_t(\phi)$, where $\phi(0) = x$, $\phi(t) = y$ and $\phi(s) \in \Omega$ for all $0 \leq s \leq t$. Some heuristic results on the behavior of $\phi_\varepsilon(x)$, $q_\varepsilon(x,y,t)$ are given in [11] in case L_ε is selfadjoint with respect to $L^2(\mu)$ where μ is a measure with $d\mu/dx = \psi(x)$, ψ smooth; see also [9] where sharper results on $q_\varepsilon(x,y,t)$ (than in (8.1)) are given in case $a_{ij} = \delta_{ij}$.

2. Once the behavior of $\phi_\varepsilon(x)$ for small ε is known, the next step is to study

e behavior of the second eigenvalue of L_ε, in the case where L_ε is selfadjoint ith respect to $L^2(\mu)$), using the variational principle for eigenvalues.

3. Generalize the results of Section 7 to elliptic partial differential erators.

4. Study the behavior of the first eigenvalue of

$$\varepsilon\Delta^2 u + b\cdot\nabla u = \lambda u \quad \text{in } \Omega,$$

$$u = \frac{\partial u}{\partial \nu} = 0 \qquad \text{on } \partial\Omega$$

$\varepsilon \to 0$.

References

1] A. Devinatz, R. Ellis and A. Friedman, The asymptotic behavior of the first real eigenvalue of second order elliptic operators with a small parameter in the highest derivatives, II, Indiana Univ. Math. J., 23 (1974), 991-1011.

2] A. Devinatz and A. Friedman, The asymptotic behavior of the first eigenvalue of differential operators degenerating on the boundary, Trans. Amer. Math. Soc., to appear.

3] A. Devinatz and A. Friedman, The asymptotic behavior of the principal eigenvalue of singularly perturbed degenerate elliptic operators, to appear.

4] A. Friedman, The asymptotic behavior of the first real eigenvalue of a second order elliptic operator with a small parameter in the highest derivatives, Indiana Univ. Math. J., 22 (1973), 1005-1015.

5] A. Friedman, Small random perturbations of dynamical systems and application to parabolic equations, Indiana Univ. Math. J., 24 (1974), 533-553; Erratum, ibid, 25 (1975), 903.

6] A. Friedman, Stochastic Differential Equations and Applications, vol. 1, Academic Press, New York, 1975.

7] A. Friedman, Stochastic Differential Equations and Applications, vol. 2, Academic Press, New York, 1976.

8] R. Z. Khasminskii, On positive solutions of the equation $\mathcal{L}u + Vu = 0$, Theory of Probability and its Applications, 4 (1959), 309-318.

9] Yu. J. Kiefer, Certain results concerning small random perturbations of dynamical systems, Theory of Probability and its Applications, 19 (1974), 487-505.

0] M. A. Krasnoselskii, Positive Solutions of Operator Equations, Groningen, Nordhoff, 1964.

1] B. J. Matkowsky and Z. Schuss, The exit problems for randomly perturbed dynamical systems, to appear.

2] M. H. Protter and H. Weinberger, On the spectrum of general second order operators, Bull. Amer. Math. Soc., 72 (1966), 251-255.

3] A. D. Ventcel, On the asymptotic behavior of the greatest eigenvalue of a second order elliptic differential operator with small parameter in the highest derivatives, Soviet Math. Doklady, 13 (1972), 13-17.

[14] A. D. Ventcel, On the asymptotics of eigenvalues with elements of order $\exp[-V_{ij}/(2\varepsilon^2)]$, Soviet Math. Doklady, 13 (1972), 65-68.

[15] A. D. Ventcel, On the asymptotic behavior of the first eigenvalue of a differential operator of the second order with small parameter in the highest derivatives, Teor. Vevojat. Primen., 20 (1975), 610-613.

[16] A. D. Ventcel and M. I. Freidlin, On small random perturbations of dynamical systems, Russian Math. Surveys, 25 (1970), 1-56.

BIFURCATION NEAR FAMILIES OF SOLUTIONS

Jack K. Hale
Lefschetz Center for Dynamical Systems
Division of Applied Mathematics
Brown University
Providence, Rhode Island 02912/USA

§1. Introduction and statement of problem.

Suppose X, Z, Λ are Banach spaces, $M: X \times \Lambda \to Z$ is continuous together with its Fréchet derivatives up through order two. Many investigations in bifurcation theory are concerned with the following problem. If $M(0,0) = 0$ and $\partial M(0,0)/\partial x$ has a nontrivial null space, find all solutions of the equation

$$M(x,\lambda) = 0 \tag{1.1}$$

for (x,λ) in a neighborhood of $(0,0) \in X \times \Lambda$.

If $\dim \Lambda = 1$; that is, there is only one parameter involved, then the existence of more than one solution in a neighborhood of zero can be proved by making assumptions only about $\partial M(0,0)/\partial x$ and $\partial M(0,0)/\partial x \partial \lambda$. However, if $\dim \Lambda \geq 2$, then the problem is much more difficult and more detailed information is needed about the function M. A careful examination of the existing literature for $\dim \Lambda \geq 2$ reveals that the additional conditions imposed on M imply, in particular, that the solution $x = 0$ of the equation

$$M(x,0) = 0 \tag{1.2}$$

is isolated (see, for example, [1], [2] and the papers on catastrophe theory in [3]). These hypotheses eliminate the possibility that Equation (1.2) has a family of solutions containing $x = 0$. Such a situation occurs, for example, for $M(x,\lambda) = Ax + N(x,\lambda)$, where A is linear with a nontrivial null space and $N(x,0) = 0$ for all x. There also are interesting applications where Equation (1.2) is nonlinear and there exists a family of solutions. For example, Equation (1.2) could be an autonomous ordinary differential equation with a nonconstant periodic orbit of period 2π with the family of solutions being obtained by a phase shift. When the differential equation in the latter situation is a Hamiltonian system, the parameters (λ_1, λ_2) could correspond to a small damping term and a small forcing term of period 2π. To the author's knowledge, the first complete investigations of special problems of each of these latter types are contained in [4], [5].

It is the purpose of this paper to begin the investigation of the abstract problem for Equation (1.1), especially to extend the results in [5]. More specifically, suppose Equation (1.2) has a one parameter family of solutions $x = p(t)$, $0 \leq t \leq 1$, $p(0) = p(1)$, which is continuous together with derivatives up through order two with $p'(t) = dp(t)/dt \neq 0$, $p'(0) = p'(1)$, $p''(0) = p'(1)$. Boundary conditions on are imposed only to avoid a special discussion at the end points of the curve defined by p. Since $M(p(t),0) = 0$ for all t, it follows that $p'(t)$ is a nonzero element of the null space $\mathfrak{N}(A(t))$ of the linear operator

$$A(t) = \partial M(p(t),0)/\partial x \tag{1.3}$$

for $0 \leq t \leq 1$.

If $\Gamma = \{p(t),\ 0 \leq t \leq 1\} \subseteq X$, the problem is to characterize the solutions of Equation (1.1) in a neighborhood of $\Gamma \times \{0\} \subseteq X \times \Lambda$. Suppose $\mathscr{R}(A(t))$ is the range of $A(t)$. For the case in which $\dim \mathfrak{N}(A(t)) = 1 = \operatorname{codim} \mathscr{R}(A(t))$, and $\Lambda = \mathbb{R}^2$, we give a solution to this problem under certain hypotheses on $\partial M(p(t),0)/\partial\lambda$. One important implication of the results can easily be stated. Suppose $\gamma \subseteq \Lambda = \mathbb{R}^2$ is a continuous curve, $0 \notin \gamma$, $0 \in \mathrm{Cl}\,\gamma$, the closure of γ, and suppose $x(\lambda)$ is a solution of system (1.1) defined and continuous for $\lambda \in \gamma$. If the set $x(\gamma) \subseteq X$ remains in a sufficiently small neighborhood of Γ and the set $x(\gamma)$ is precompact, then all limit points of $x(\lambda)$ as $\lambda \in \gamma$ approaches zero belong to Γ, but $x(\lambda)$ has a limit as $\lambda \to 0$ if and only if $\cot^{-1}(\lambda_1/\lambda_2)$ approaches a constant as $\lambda \in \gamma$ approaches zero.

§2. Statement and implications of results.

For any Banach spaces X,Z, we let $C^k(X,Z)$ be the linear space of all functions from X to Z which are continuous together with all derivatives up through order k. If no confusion may arise, we sometimes write C^k for $C^k(X,Z)$. For any finite collection of elements $q_1,\ldots,q_k$ of a Banach space, we let $[q_1,\ldots,q_k]$ denote the linear subspace spanned by $q_1,\ldots,q_k$. By our boundary condition on p, we may suppose $p \in C^2(\mathbb{R},X)$ and is 1-periodic, that is, periodic of period 1. Suppose $[p'(t)] = \mathfrak{N}(A(t))$ and there is a $q \in C^2(\mathbb{R},Z)$ 1-periodic, such that $[q(t)] \oplus \mathscr{R}(A(t)) = Z$. If $\mathscr{B}(X)$ is the Banach space of bounded linear operators on X, let $U \in C^2(\mathbb{R}, \mathscr{B}(X))$ be such that $U(t)$ is a projection onto $\mathfrak{N}(A(t))$ and let $E \in C^2(\mathbb{R}, \mathscr{B}(Z))$, $E(t)$ a projection onto $\mathscr{R}(A(t))$, $I - E(t)$ a projection onto $[q(t)]$. Also suppose U,E are 1-periodic.

If $\Lambda = \mathbb{R}^2$, $\lambda = (\lambda_1,\lambda_2) \in \Lambda$, define $\alpha_j \in C^2(\mathbb{R},\mathbb{R})$, $j = 1,2$,

1-periodic, by the relation

$$\alpha_j(t)q(t) = (I-E(t))\partial M(p(t),0)/\partial\lambda_j \qquad (2.1)$$

Our first hypothesis on $\alpha(t) = (\alpha_1(t),\alpha_2(t))$ is

(H_1) $\qquad \alpha(t) \neq 0$ for $t \in \mathbb{R}$.

If $\beta(t) = (\alpha_2(t),-\alpha_1(t))$, then $\beta(t) \neq 0$ by Hypothesis (H_1) and we can let $\phi(t)$ be the angle measured in the counterclockwise direction which $\beta(t)$ makes with the horizontal axis. The function $\phi \in C^2(\mathbb{R},\mathbb{R})$ and is 1-periodic. We impose the following hypotheses on ϕ:

(H_2) The function $\phi'(t)$ has at most a finite set of zeros $\{t^k, k = 1,2,\ldots,n\} \subseteq [0,1)$ and $\phi''(t^k) \neq 0$ for $k = 1,2,\ldots,n$.

(H_3) $\phi(t^j) \neq \phi(t^k)$, $j \neq k$, $j,k = 1,2,\ldots,n$.

We now state the main results of the paper together with implications. The proofs will be given in Section 3. Suppose γ is a smooth curve in $\mathbb{R}^2$ through the origin. If for any $q \in \gamma$, $q \neq (0,0)$, L_q^1 denotes a positively oriented normal to L_q at q, we say γ is crossed from right to left at q if γ is crossed by moving along L_q^1 in the positive direction.

Theorem 2.1. If Hypotheses (H_1)-(H_3) are satisfied, then there exist neighborhoods U of Γ, V of $\lambda = (0,0)$, and $s_0 > 0$, such that, for each $t^j \in \{t^k, k = 1,2,\ldots,n\}$, there corresponds a unique curve $\mathcal{C}_j \subseteq V$, tangent to the line $\alpha(t^j)\cdot\lambda = 0$ at zero, $\mathcal{C}_j \cap \partial V \neq \phi$, each $\mathcal{C}_j$ intersects lines through the origin in at most one nonzero point, these curves intersect only at $(0,0)$, the number of solutions of (1.1) in U increases (or decreases) by exactly two as $\mathcal{C}_j$ is crossed from right to left if t^j is a relative minimum (or maximum) of ϕ.

The curves $\mathcal{C}_j$ can be defined parametrically in the form $\lambda = s\beta_j(s)$, $0 \leq s < s_0$ where $\beta_j \in C^2([0,s_0),\mathbb{R}^2)$, $|\beta_j(s)| = 1$, $0 \leq s < s_0$, and $\alpha(t^j)\cdot\beta_j(0) = 0$. If t_*,t^* are the absolute minimum, maximum, respectively, of ϕ and

$$S(V) = \{\lambda \in V\colon \lambda\cdot\beta^*(s) < 0 < \lambda\cdot\beta_*(s),\ 0 \leq s < s_0\},$$

where $\lambda = s\beta^*(s)$, $\lambda = s\beta_*(s)$ are the curves corresponding to t^* and t_*, then there are no solutions of Equation (1.1) in U for $\lambda \in S(V)$, at least two in $S^c(V) = V\backslash S(V)$ and all solutions are distinct in the interior of $S^c(V)$.

The curves $\mathscr{C}_j$ in Theorem 2.1 are called the <u>bifurcation curves</u>. To see how easy it is to obtain the complete qualitative picture of the bifurcations near $\lambda = 0$, let us consider a few special cases. I $\phi(t)$ has only one maximum at t^* and one minimum at t_*, there are only two bifurcation curves $\mathscr{C}_*, \mathscr{C}^*$ corresponding to t_*, t^*, respectively. There are no solutions of Equation (1.1) in U for $\lambda \in S(V)$ and exactly two solutions in $S^c(V)$ which are distinct in the interior of $S^c(V)$ (see Figure 1). If there are two maxima and two minima (there must always be an even number of maxima and minima by periodicity), then the situation is depicted in Figure 2. By changing the function ϕ, one can obtain every possible rotation of these pictures.

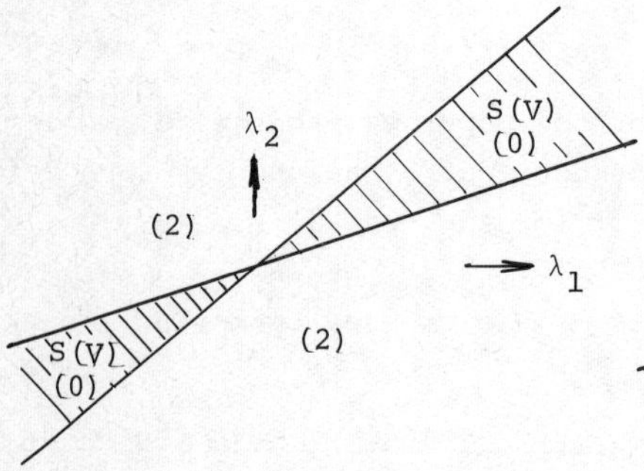

Figure 1.

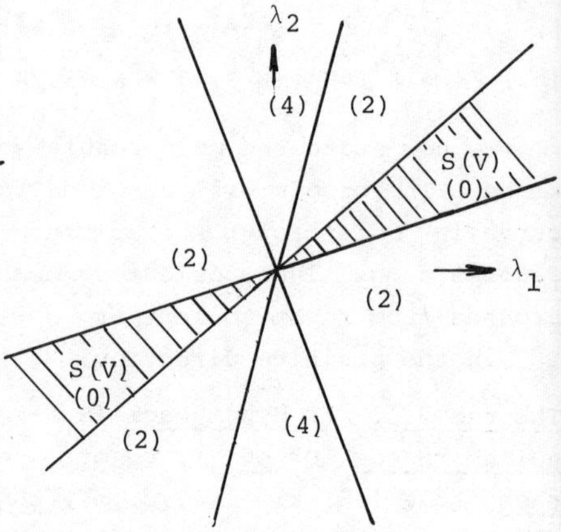

Figure 2.

Another interesting special case is $\alpha_1(t) = -1$, $t \in \mathbb{R}$. Hypothes (H_1) is always satisfied, $\beta(t) = (\alpha_2(t), 1)$, $\phi(t) = \cot^{-1}\alpha_2(t)$ and th hypotheses (H_1), (H_2) are equivalent to

(H_2') The function α_2' has a most finite set of zeros $\{t^k, k = 1,2,\ldots,n\} \subseteq [0,1)$ and $\alpha_2''(t^k) \neq 0$, $k = 1,2,\ldots,n$.

(H_3') $\alpha_2(t^j) \neq \alpha_2(t^k)$, $j \neq k$, $j, k = 1,2,\ldots,n$.

Theorem 2.1 for this case is essentially contained in [5]. Since $\alpha_1(t) = -1$, the set $S(V)$ must contain the λ_1-axis and, thus, the bifurcation diagram is a rotated version of the ones in Figures 1 and

The following result gives some information about the possible behavior of the solutions of Equation (1.1) as $\lambda \to 0$.

<u>Theorem 2.2</u>. <u>Suppose Hypotheses</u> (H_1)-(H_3) <u>are satisfied</u>, U, V <u>are th</u>

neighborhoods given in Theorem 2.1, and suppose γ *is a continuous curve defined parametrically by* $\lambda_1 = \lambda_1(\tau)$, $\lambda_2 = \lambda_2(\tau)$, $0 \le \tau \le 1$, *and* $\lambda_1^2(\tau) + \lambda_2^2(\tau) = 0$ *if and only if* $\tau = 0$. *Also, suppose* $\gamma \subseteq V$ *and for each point* $(\lambda_1(\tau), \lambda_2(\tau)) \in \gamma$, *there is a solution* $x(\tau) \in U$ *of Equation* (1.1) *which is continuous in* τ *on the half open interval* $(0,1]$. *If*

$$\mathscr{S}(\gamma) = \{x(\tau), \quad 0 < \tau \le 1\} \subseteq X \tag{2.2}$$

is precompact, and

$$\begin{aligned} \phi_m(\gamma) &= \liminf_{\tau\to 0} \cot^{-1}(\lambda_1(\tau)/\lambda_2(\tau)) \\ \phi_M(\gamma) &= \limsup_{\tau\to 0} \cot^{-1}(\lambda_1(\tau)/\lambda_2(\tau)) \end{aligned} \tag{2.3}$$

then there is an interval $I(\gamma) \subseteq [0,1]$ *such that* $\phi(I(\gamma)) = [\phi_m(\gamma), \phi_M(\gamma)]$ *and*

$$(\mathrm{Cl}\,\mathscr{S}(\gamma))\backslash\mathscr{S}(\gamma) = \{p(t), \quad t \in I(\gamma)\}. \tag{2.4}$$

A consequence of the above result is the following

Corollary 2.1. If $\gamma, x(\tau)$ *satisfy the conditions of Theorem 2.2, then a necessary and sufficient condition that* $x(\tau)$ *have a limit as* $\tau \to 0$ *is that* $\cot^{-1}(\lambda_1(\tau)/\lambda_2(\tau))$ *has a limit* ϕ_0 *as* $\tau \to 0$. *In this case,* $x(\tau) \to p(t_0)$ *where* $t_0 \in [0,1)$ *is a solution of the equation* $\phi(t) = \phi_0$.

The fact that one can obtain solutions which are not continuous in λ at $\lambda = 0$ is not surprising. Consider the scalar equation $\lambda_1 x - \lambda_2 = 0$ which has the solution $x = \lambda_2/\lambda_1$ for $\lambda_1 \ne 0$. Along a curve $\gamma \in \mathbb{R}^2$, this solution has a limit as $\lambda \to 0$ in γ if and only if λ_1/λ_2 approaches a limit as $\lambda \to 0$ in γ.

Let us now make a more interesting application to the second order scalar ordinary differential equation

$$\frac{d^2x}{ds^2} + g(x) + \lambda_1 h(s)\frac{dx}{ds} - \lambda_2 f(s) = 0 \tag{2.5}$$

where h, f are continuous and 1-periodic, $g \in C^2(\mathbb{R},\mathbb{R})$, $xg(x) > 0$ for $x \ne 0$. For $\lambda_1 = \lambda_2 = 0$, the equation

$$\frac{d^2x}{ds^2} + g(x) = 0 \tag{2.6}$$

has a general solution of the form $x = \psi(\omega(a)s + t, a)$, $(a,t) \in \mathbb{R}^2$, where $\psi(\zeta,a) = \psi(\zeta+1,a)$ for all (ζ,a), and $(a,0) = (x(0), dx(0)/ds)$. We suppose Equation (2.6) has a nondegenerate 1-periodic orbit; that is,

There is an $a_0 > 0$ *such that* $\omega(a_0) = 1$, $d\omega(a_0)/da \ne 0$. (2.7)

Let $Z = \{y: \mathbb{R} \to \mathbb{R}$ which are continuous and 1-periodic$\}$ and use the supremum norm on Z. Let $X = \{y \in Z: y$ has continuous derivatives up through order two$\}$ and use the usual C^2 norm on X. If we define $M: X \times \Lambda \to Z$, $\Lambda = \mathbb{R}^2$ by

$$M(x,\lambda)(s) = \frac{d^2x(s)}{ds^2} + g(x(s)) + \lambda_1 h(s)\frac{dx(s)}{ds} - \lambda_2 f(s)$$

then we are in a position to apply the previous results. In fact, if $p(t)(s) = \psi(\omega(a_0)s + t, a_0)$, then $p(t) \in X$ and satisfies $M(p(t),0) = 0$, $0 \leq t \leq 1$. Also, Hypothesis (2.7) implies that $\dim \mathfrak{N}(A(t)) = 1 = \operatorname{codim} \mathscr{R}(A(t))$, where $A(t) = \partial M(p(t),0)/\partial x$. Furthermore, the function $\dot{p}(t)$ is a basis for $\mathfrak{N}(A(t))$ and a complement for $\mathscr{R}(A(t))$. It is now an obvious calculation to see that the functions $\alpha_1(t), \alpha_2(t)$ in (2.1) are given by

$$\alpha_1(t) = -\int_0^1 h(s)\dot{p}(s+t)^2 ds, \quad \alpha_2(t) = \int_0^1 \dot{p}(s+t)f(s)ds. \tag{2.8}$$

If (α_1, α_2) satisfy (H_1)-(H_3), then the above results are directly applicable to the determination of the bifurcation curves for the 1-periodic solutions of Equation (2.5) which lie in a neighborhood of the periodic orbit $\Gamma \subseteq \mathbb{R}^2$ of Equation (2.6) defined by $\Gamma = \{(p(s), dp(s)/ds), 0 \leq s \leq 1$. For $h(s) = 1$, $0 \leq s \leq 1$, these results were previously obtained in [5]. A detailed explanation of the manner in which the 1-periodic solutions wind onto the cylinder $\Gamma \times \mathbb{R}$ as $\lambda \to 0$ along a curve $\gamma \in \mathbb{R}^2$ is given in [5]. Also, a reasonable physical explanation for the discontinuities of the solutions at $\lambda = 0$ is given in [5].

§3. Proof of the results.

Our objective in this section is to give the essential elements of the proofs of the results of Section 2. The notation of that section will be used without explanation.

By the Implicit Function Theorem and the compactness of Γ, one obtains the following result.

Lemma 3.1. There is a $\delta > 0$ such that the transformation $x \mapsto (t,y)$

$$x = p(t) + y, \quad y \in [I-U(t)]X, \tag{3.1}$$

from a neighborhood of Γ to $[0,1) \times (I-U(t))X$ is a diffeomorphism for $t \in [0,1)$, $|y| < \delta$.

For the determination of all solutions of Equation (1.1) in a sufficiently small neighborhood of Γ, Lemma 3.1 implies that it is sufficient to consider x given by Equation (3.1) and $|y|$ in a sufficiently small neighborhood of zero. If x is a solution of

Equation (1.1) and y is defined by Equation (3.1), then y satisfies the equation

$$0 = M(p(t) + y,\lambda) \stackrel{\text{def}}{=} A(t)y + N(t,y,\lambda) \tag{3.2}$$

where $N(t,y,\lambda) = M(p(t) + y,\lambda) - A(t)y$. By the boundary conditions on p, we may consider this equation for $t \in R$. Decomposing this equation into its components in $E(t)Z$ and $[I-E(t)]Z$ and using the fact that $A(t)$ as a mapping from $[I-U(t)]X$ onto $E(t)Z$ has a bounded inverse (this is the method of Liapunov-Schmidt), there exist $\lambda_0 > 0$, $\delta > 0$ and a unique function $y^* \in C^2(\mathbb{R} \times \{|\lambda| < \lambda_0\}$, $(I-U(t))X)$, $y^*(t,0) = 0$ for all t, such that Equation (3.2) has a solution for $t \in \mathbb{R}$, $|y| < \delta$, $|\lambda| < \lambda_0$ if and only if $y = y^*(t,\lambda)$ and (t,λ) satisfies the bifurcation equation

$$0 = F(t,\lambda) \stackrel{\text{def}}{=} (I-E(t))[M(p(t) + y^*(t,\lambda),\lambda) - A(t)y^*(t,\lambda)] \tag{3.3}$$

If we define the scalar function $f(t,\lambda)$ by the relation

$$F(t,\lambda) \stackrel{\text{def}}{=} f(t,\lambda)q(t) \tag{3.4}$$

then the bifurcation equation is equivalent to the scalar equation

$$f(t,\lambda) = 0 \tag{3.5}$$

for $t \in \mathbb{R}$, $|\lambda| < \lambda_0$. Since $y^*(t,0) = 0$ for all t, it follows that

$$f(t,0) = 0, \quad t \in \mathbb{R}. \tag{3.6}$$

Relation (3.6) reflects the fact that Equation (1.2) does not have an isolated solution. Equality (3.6) is the basic reason why this problem differs from the usual bifurcation problem.

If $\Lambda = R^2$, $\lambda = (\lambda_1,\lambda_2) \in \mathbb{R}^2$, then

$$\partial f(t,0)/\partial\lambda_j = \alpha_j(t), \quad j = 1,2 \tag{3.7}$$

where each α_j, $j = 1,2$, is defined in Equation (2.1). The function f can thus be written as

$$f(t,\lambda) = \alpha(t)\cdot\lambda + h(t,\lambda)$$

where $h(t,0) = 0$, $\partial h(t,0)/\partial\lambda = 0$. For any $\lambda \neq 0$, solving Equation (3.5) is equivalent to solving the equation

$$\alpha(t)\cdot(\lambda/|\lambda|) + h(t,\lambda)/|\lambda| = 0. \tag{3.8}$$

If $\lambda/|\lambda| = \beta$, $H(t,\beta,|\lambda|) = h(t,\beta|\lambda|)/|\lambda|$, then $\beta \in S^1 = \{\beta \in \mathbb{R}^2$: $|\beta| = 1\}$ and H is C^2 in its arguments. The discussion of the solutions of Equation (3.8) becomes equivalent to the discussion of the equation

$$G(t,\beta,s) \stackrel{\text{def}}{=} \alpha(t)\cdot\beta + H(t,\beta,s) = 0 \tag{3.9}$$

for $t \in [0,1)$, $\beta \in S^1$, s small and nonnegative. In the following, we always understand $\beta \in S^1$ even though it may not be said explicit[ly]

If $\alpha(t_0)\cdot\beta_0 = 0$, then $G(t_0,\beta_0,0) = 0$. If $\alpha'(t_0)\cdot\beta_0 \neq 0$, then $\partial G(t_0,\beta_0,0)/\partial t \neq 0$ and the Implicit Function Theorem implies there is an $s_0 = s_0(t_0,\beta_0) > 0$ and a unique solution $t^*(\beta,s)$ of Equation (3.9) for $|\beta-\beta_0| < s_0$, $0 \leq s < s_0$, $t^*(\beta_0,0) = t_0$.

To complete the proof, we need to reformulate Hypotheses (H_2), (H_3) in an equivalent form. The vector $\alpha(t)$ defines a continuous linear functional on $\mathbb{R}^2$ by the relation $\alpha(t)\cdot\lambda$, $\lambda \in \mathbb{R}^2$. If the null space of $\alpha(t)$ is denoted by $\mathfrak{N}(\alpha(t))$, then $(\alpha_2(t),-\alpha_1(t))$ is a basis for $\mathfrak{N}(\alpha(t))$. By computing ϕ',ϕ'', one easily observes that (H_2), $(H_3$[)] are equivalent to

(H_2) The vector $\alpha'(t) \in \mathbb{R}^2$ is orthogonal to $\mathfrak{N}(\alpha(t))$ at most at a finite number of points $\{t^k, k = 1,2,\ldots,n\} \subseteq [0,1)$ and $\alpha''(t^k)$ is not orthogonal to $\mathfrak{N}(\alpha(t^k))$ for any $k = 1,2,\ldots,n$

(H_3) The lines through the origin and $\alpha(t^j)$ and $\alpha(t^k)$ are not colinear for $j \neq k$, $j,k = 1,2,\ldots,n$.

The numbers t^k here are the same as before.

If $\alpha'(t_0)\cdot\beta_0 = 0$, then Hypothesis (H_2) implies $\alpha''(t_0)\cdot\beta_0 \neq 0$. Thus, the Implicit Function Theorem implies there is an $s_0 = s_0(t_0,\beta_0) > 0$ and a function $t^*(\beta,s)$, $t^*(\beta_0,0) = t_0$, such that

$$\partial G(t^*(\beta,s),\beta,s)/\partial t = 0$$

for $|\beta-\beta_0| < s_0$, $0 \leq s < s_0$ and $t^*(\beta,s)$ is unique in the region $|t-t_0| < s_0$. Thus, the function $Q(\beta,s) \overset{\text{def}}{=} G(t^*(\beta,s),\beta,s)$ is a maximum or minimum of $G(t,\beta,s)$ with respect to t at (β,s). A few elementary calculation show that $Q(\beta_0,0) = 0$, $\partial Q(\beta_0,0)/\partial\beta = \alpha(t_0)$. Therefore, the derivative of $Q(\beta,s)$ with respect to β on the spher[e] S^1 at $\beta = \beta_0$, $s = 0$ is the inner product of the vector $\alpha(t_0)$ with a unit vector orthogonal to β_0. But this vector will be a nonzero constant times $\alpha(t_0)$ since $\alpha(t_0) \neq 0$ by Hypothesis (H_1). Consequently, $\partial Q(\beta_0,0)/\partial\beta$ on S^1 is nonzero. The Implicit Function Theor[em] implies there is a $\delta(\beta_0) > 0$ and a function $\beta^*(s) \in S^1$, $\beta^*(0) = \beta_0$, such that $Q(\beta^*(s),s) = 0$ for $0 \leq s < \delta(\beta_0)$ describes a curve. On one side of this curve, there are two simple solutions of Equation (3.9) and no solutions on the other side. In terms of the original coordinates (λ_1,λ_2), this implies there are two solutions of Equation (3.5) near t_0 on one side of the curve $\lambda = s\beta^*(s)$, $0 \leq s < \delta(\beta_0)$ and none on the other. This curve in λ-space is a bifurcation curve and is tangent to the line $\alpha(t_0)\cdot\lambda = 0$ at $\lambda = 0$. The fact that the number of solutions increases or decreases as stated in the theorem is clear.

The above analysis can be applied to each of the points t^j in Hypothesis (H_2) to obtain an $s_0 > 0$ such that all solutions of Equation (3.9) for $|t-t^j| < s_0$, $\beta \in S^1$, $0 \leq s < s_0$ are determined by the argument above. We obtain the bifurcation curves as well. The complement of the intervals $|t-t^j| < s_0$, $j = 1,2,\ldots,n$, in $[0,1)$ is compact and $\alpha'(t_0)\cdot\beta_0 \neq 0$ for any t_0,β_0 satisfying $\alpha(t_0)\cdot\beta_0 = 0$. A repeated application of the Implicit Function Theorem shows one can choose s_0 so that no further bifurcations occur in this complement for any $\beta \in S^1$, $0 \leq s < s_0$. Returning to the original coordinates (λ_1,λ_2), we see that the complete bifurcation diagram has been obtained for a full neighborhood of $\lambda = 0$.

To describe precisely the bifurcation pattern as stated in Theorem 2.1, we need to know that no two bifurcation curves obtained by the above process coincide. This is the only reason Hypothesis (H_3) is imposed. This proves the first part of Theorem 2.1.

The last part of the theorem is clear from the definitions of the terms involved. This completes the proof of Theorem 2.1.

To prove Theorem 2.2, we first note that the method of Liapunov-Schmidt implies there is a $\tau_0 > 0$ and a continuous function $t(\tau) \subseteq [0,1]$, $0 < \tau < \tau_0$ such that the solution $x(\tau)$ is given by

$$x(\tau) = p(t(\tau)) + y^*(t(\tau),\lambda(\tau)), \quad 0 < \tau < \tau_0,$$

where $y^*(t,\lambda)$ is the solution of Equation (3.2), $(t(\tau),\lambda(\tau))$ satisfy Equation (3.5) or, equivalently, Equation (3.8). Suppose there is a sequence $\tau_j \to 0$ such that $\cot^{-1}(\lambda_1(\tau_j)/\lambda_2(\tau_j)) \to \phi_0 \in [0,2\pi]$ as $j \to \infty$, or, equivalently, $\lambda(\tau_j)/|\lambda(\tau_j)| \to \beta_0 \in S_1$ as $j \to \infty$. Without loss of generality, we may assume $t(\tau_j) \to t_0 \in [0,1]$ as $j \to \infty$. Then $x(\tau_j) \to p(t_0)$, $\alpha(t_0)\cdot\beta_0 = 0$, $\phi(t_0) = \phi_0$. Since all functions are continuous, the conclusion of Theorem 2.2 follows immediately, and the proof is complete.

To prove Corollary 2.1, suppose the conditions of Theorem 2.2 are satisfied and the interval $[\phi_m(\gamma),\phi_M(\gamma)]$ consists of more than one point, then $x(\tau)$ cannot have a limit as $\tau \to 0$ although every limit point satisfies Equation (1.2). If $x(\tau)$ has a limit as $\tau \to 0$, then $\mathcal{S}(\gamma)$ is precompact and it is, therefore, necessary that $\cot^{-1}(\lambda_1(\tau)/\tau_2(\tau))$ approach a limit as $\tau \to 0$. Conversely, if $\cot^{-1}(\lambda_1(\tau)/\lambda_2(\tau)) \to \phi_0$ as $\tau \to 0$, then $\lambda(\tau)/|\lambda(\tau)| \to \beta_0 \in S^1$ as $\tau \to 0$ and $\alpha(t(\tau))\cdot\beta_0 \to 0$ as $\tau \to 0$ from the argument used in the proof of Theorem 2.2. Hypothesis (H_2) implies the set of $t \in [0,1]$ such that $\alpha(t)\cdot\beta_0 = 0$ is isolated. Since $t(\tau)$ is continuous for $0 < \tau \leq 1$, this implies $t(\tau) \to t_0 \in \{t^k, k = 1,2,\ldots,n\}$ as $\tau \to 0$. The argument used in the proof of Theorem 2.2 implies $x(\tau) \to p(t_0)$ as $t \to \infty$ and the proof of Corollary 2.1 is complete.

REFERENCES

[1] Chow, S., Hale, J.K. and Mallet-Paret, Applications of generic bifurcation I, II. Arch. Rat. Mech. Ana. 59(1975), 159-188; 62(1976), 209-236.

[2] Thompson, J.M.T. and Hunt, G.W., A General Theory of Elastic Stability. Wiley, 1973.

[3] Dynamical Systems - Warwick 1974, Lecture Notes in Math., vol. 468, Springer-Verlag.

[4] Hale, J.K. and Rodrigues, H.M., Bifurcation in the Duffing equation with independent parameters I, II. Proc. Royal Soc. Edinburg, Ser. A. To appear.

[5] Hale, J.K. and Táboas, P., Effect of damping and forcing in a second order equation. J. Nonlinear Analysis - Theory, Method, Applications (JNA-TMA). To appear.

This research was supported in part by the Air Force Office of Scientific Research under AFOSR 71-2078D, in part by the National Science Foundation under MPS 71-02923 and in part by the U.S. Army Research Office under AROD AAG 29-76-6-0052.

NON-LINEAR DIFFERENTIAL EQUATIONS; QUESTIONS AND SOME ANSWERS

By EINAR HILLE

Since 1969 I have studied non-linear differential equations and here is a brief account of results obtained by me and by other writers. The results belong to five different but partly overlapping domains.

1. Singularities of solutions. In the first order case

$$P(z, w, w') = 0 \tag{1.1}$$

let P be a polynomial in the variables. The determinateness theorem of Paul Painlevé shows that the movable singularities are algebraic. Their nature can be obtained by substituting a test-power $a(z - z_0)^{-\alpha}$ for w in (1.1) and balancing infinitudes.

Second order DE's present totally different possibilities. Thus

$$z w'' = w^2 \tag{1.2}$$

mounts fantastic singularities. At the fixed singularity $z = 0$ there are besides solutions with simple poles also logarithmic psi-series

$$P_0 + \sum_{n=1}^{\infty} P_n(\log z) z^n \tag{1.3}$$

where P_0 is a constant while $P_n(t)$ is a polynomial in t of degree $[\frac{1}{2}(n + 1)]$ involving P_0 and another arbitrary constant. Such psi-series occur at all movable infinitudes where there are expansions

$$w(z) = \sum_{n=0}^{\infty} P_n[\log(z_0 - z)](z_0 - z)^{n-2}. \tag{1.4}$$

Here $P_n(t)$ is a polynomial in t of degree $[n/6]$. The solution has a pseudo-pole at $z = z_0$ and the logarithmic perturbation appears from the seventh term on.

Equation (1.2) is an Emden equation [5]

$$w'' = z^{1-m} w^m \tag{1.5}$$

for the special case $m = 2$. The equation introduced by R. Emden in 1907 with applications to cosmology and stellar dynamics differs from (1.5) by having a minus sign in the left member; (1.5) is more interesting to the mathematician. For

$m = \frac{3}{2}$ we get an equation found in 1927 by L. H. Thomas [15] and E. Fermi [6] in a study of electron clouds around the nucleus of an atom. I have devoted several papers to these equations, see e.g. [8]. Real-valued positive solutions normally have a graph with a vertical asymptote beyond which the solution ceases to be real.

It was shown by W. A. Coppel [4] in 1966 that (1.5) is reducible to the quadratic system

$$\dot{x} = x(1 - x + y), \quad \dot{y} = y(2 - m + mx - y) \tag{1.6}$$

via the substitution

$$x = z\,\frac{w'}{w}, \quad y = \frac{z^{1-m}w^m}{w'}, \quad t = \log z. \tag{1.7}$$

I have discussed the analytical nature of the singularities of solutions of quadratic systems [9]. Using different methods Russell Smith [14] in 1975 discussed such problems for general polynomial systems. Extensions to higher dimensions are desirable.

2. Rate of growth and order of solutions. Growth problems in the real domain have been studied by me for the Emden-Fermi-Thomas equations. Thus a solution $w(x)$ defined at $x = a > 0$ with $0 < w(a) = b$, $0 \leqq w'(a) = c$ becomes infinite as $6(x_0 - x)^{-2}$ as x increases to a value x_0 where $a < x_0 < X_0$ and X_0 is a simple function of a and b provided $3c^2 \geqq 2b^3a^{-1}$. The first step in the proof is to derive the inequality

$$[w'(x)]^2 > \frac{2}{3}x^{-1}[w(x)]^3 \tag{2.1}$$

from which the desired inequalities follow. Such phenomena arise for all Emden equations and also for more general classes of DE's.

More general growth questions arise in the complex domain and if the solution is entire or meromorphic the Nevanlinna value distribution theory becomes useful. Using the Wiman-Valiron theory of entire functions G. Valiron proved in 1923 that in the first order case if a solution is an entire function the order is necessarily finite. This is no longer true for the second order case; thus $e^{\sin z}$ satisfies a DE of the form $P(w, w', w'') = 0$. For meromorphic solutions Kôsaku Yosida in 1933 used Nevanlinna theory to discuss Riccati equations

(2.2) $$Q(z)w' = P_0(z) + P_1(z)w(z) + P_2(z)[w(z)]^2$$

and hyper-Riccati equations

(2.3) $$Q(z)[w'(z)]^n = P_0(z) + P_1(z)w(z) + \ldots + P_{2n}(z)[w(z)]^{2n}.$$

Here $P_j(z)$ is a polynomial of degree δ_j and $Q(z)$ one of degree q. Suppose that the equation has at least one transcendental meromorphic solution and set $\max \delta_j = p$. Then Yosida found that

(2.4) $$T(r, w) = \begin{cases} O[r^{2(p-q)/n+2}] \\ O(\log^2 r) \end{cases}$$

according as $2(p-q)/n \neq -2$ or $= -2$. He remarks that if the coefficients are constants so that the equation is of Briot-Bouquet type then $T(r, w) = O(r^2)$. Since the solution may be an elliptic function the estimate is the best possible.

More general results were found by A. A. Gol'dberg [7] in 1956. Most of his results apply to equations of the form

(2.5) $$(w')^n + P_1(z, w)(w')^{n-1} + \ldots + P_n(z, w) = 0$$

(2.6) $$P_j(z, w) = \sum_k a_{jk}(z)w^k, \quad 0 < j \leq n.$$

Here the a_{jk} are rational functions of z of degree δ_{jk} at ∞. Let

(2.7) $$P = \max_{1 \leq j \leq n} \{\frac{1}{j} \max_k \delta_{jk}\}.$$

If $w(z)$ is a single-valued solution of (2.5) and if

i) $p < -1$, then $T(r, w) = O(\log r)$ and $w(z)$ is rational;

ii) $p = -1$, then $T(r, w) = O(\log^2 r)$;

iii) $p > -1$, then $T(r, w) = O(r^{2p+2})$.

Gol'dberg's main tool is the Ahlfors-Shimizu characteristic with estimates of the spherical area function $A(r, w^m)$ for $m \geq 1$.

Both Gol'dberg and Yosida found $O(\log^2 r)$ as the lowest order for $T(r, w)$ when $w(z)$ is a transcendental meromorphic solution. Steven Bank and Robert Kaufman [2] have shown that this order is actually attained by the meromorphic function

(2.8) $$\wp\{\log[z + (z^2 - 1)^{\frac{1}{2}}] ; \tfrac{1}{2}, \pi i\}$$

which satisfies

(2.9) $$(z^2 - 1)(w')^2 = 4\, w^3 - g_2 w - g_3$$

for an appropriate choice of g_2 and g_3. They ask if $O(\log^2 r)$ is the minimal order. This is not clear from the results of Gol'dberg and Yosida. There is a corresponding Bank-Kaufman minimal problem for every order of the DE's; since the Weierstrass $\wp$-function satisfies DE's of every given order, the minimal rate of growth cannot exceed $O(\log^2 r)$. Bank and Kaufman have also extended Gol'dberg's theorem to the case where the $P_j(z, w)$ are polynomials in z and meromorphic in w. Bank has produced a number of papers dealing with growth and order questions of first and second order equations under different assumptions on the coefficients See e.g. [1]. See also C-C Yang [17].

3. <u>The Malmquist theorem</u>. J. Malmquist [13] proved that if

(3.1) $$w' = R(z, w), \ R \text{ rational in } z \text{ and } w$$

has a meromorphic solution $w(z)$ then either $w(z)$ is a rational function of z or the DE is a Riccati equation (2.2). Using the Nevanlinna theory Yosida [18,19] gave a new proof of this theorem in 1933 and also extended the investigation to the case

(3.2) $$(w')^n = R(z, w)$$

where the existence of a transcendental meromorphic solution implies that the DE is hyper-Riccati (2.3). A systematic scrutiny of the implications of the Nevanlinna theory for DE's was started by Hans Wittich some 30 odd years ago (see [16]) and this led him also to a new proof of the Malmquist theorem. Wittich stressed the finiteness of the Nevanlinna order of meromorphic solutions and for this I gave a new proof [10]. The assumption that R is rational in z is unnecessarily restrictive. In recent years Ilpo Laine [12] and Chung-Chun Yang [17] assume instead that the coefficients $C_j(z)$ are entire or meromorphic and that the solution $w(z)$ is <u>admissible</u> in the sense that

(3.3) $$T(r, C_j) = o[T(r, w)], \ \forall\, j.$$

f this holds the Malmquist-Yosida conclusions are valid.

In [11] I have attacked these problems along different lines. It is not enough to assume that the solution has infinitely many poles, but one can manage if the frequency function for the poles grows fast enough. More precisely, suppose that

(3.4) $$N(r, \infty; w) > K \log C(r) \quad \text{or} \quad N(r, \infty; w) > L\, D(r)$$

according as the coefficients $C_j(z)$ are entire or meromorphic. Here $C(r)$ is a common majorant of the C_j's and $D(r)$ is a majorant of $T[r, C_j(z)]$. If $K > 7$ or $L > 2$, the DE (3.1) is Riccati. Various extensions are also proved for (3.2).

It should be observed that in the meromorphic case the poles of the coefficients are fixed singularities so the existence of solutions meromorphic in the finite plane is somewhat problematic. The Lamé-Riccati equation

(3.5) $$w' = w^2 - 2\wp(z) - e_1 \quad \text{with} \quad w(z) = -\frac{1}{2}\frac{\wp'(z)}{\wp(z) - e_1}$$

shows that infinitely many fixed singularities does not exclude the existence of meromorphic solutions. On the other hand,

(3.6) $$w' = \sec^2 z[1 + w^2] \quad \text{with} \quad w(z) = \tan[\tan z]$$

gives a case where the singularities are points of condensation of poles of the solutions.

4. Polar neighborhoods. This is a concept from the theory of meromorphic functions which was applied to DE's by P. Boutroux in 1908 and revived by H. Wittich in his proof of the Malmquist theorem [16]. The crucial step is to find non-overlapping neighborhoods of the poles outside of which the solution is bounded or has a modest rate of growth. If such sets have been found then bounds for $T(r, w)$ can be found by solving a closest packing problem: if there is no overlapping, how many small disks of known diameters can be placed (i) in a given large disk or (ii) with centers on the rim thereof? The first problem leads directly to upper bounds for $N(r, \infty; w)$ while the second gives estimates of the proximity function $m(r, \infty; w)$. I have used this idea for the Malmquist-Yosida theorems, for the first two Painlevé transcendents and for Briot-Bouquet equations.

Recently Laine has also used it.

5. Briot-Bouquet equations. This name stands for DE's

$$P[w, w^{(k)}] = 0 \tag{5.1}$$

where $P(x, y)$ is a polynomial in x and y with constant coefficients. The classical case is $k = 1$ but I have recently also studied $k > 1$. For such BB-equations the interesting case is that where there are transcendental meromorph solutions. If such solutions exist the algebraic curve

$$P(x, y) = 0 \tag{5.2}$$

has the parametric representation

$$x = w(z), \quad y = w^{(k)}(z) \tag{5.3}$$

by meromorphic functions. E. Picard showed in 1887 that such a representation is possible (if and) only if the genus of the curve is 0 or 1 so this is a necessary condition for the existence of transcendental solutions of (5.1). For any value of k there exist BB-equations with meromorphic solutions for Ch. Briot and J. Cl. Bouquet proved in 1856 that all elliptic functions satisfy such DE's. This is simply a consequence of the fact, also proved by them, that two elliptic functions with the same periods are algebraic functions of each other.

Take $k = 1$ and the BB-equation

$$P_0(w)(w')^n + P_1(w)(w')^{n-1} + \ldots + P_n(w) = 0 \tag{5.4}$$

and let the degree of the polynomial $P_j(w)$ be δ_j. L. Fuchs showed in 1881 that the existence of solutions single-valued in the finite plane requires that P_0 is a constant and

$$\delta_j \leqq 2j, \ \forall\, j. \tag{5.5}$$

If w' is replaced by $w^{(k)}$ then (5.5) should be replaced by

$$\delta_j \leqq (k + 1)j. \tag{5.6}$$

For $k = 1$ Fuchs, H. Poincaré and L. Schlesinger have determined the nature of the single-valued solutions. They are rational functions of z or of e^{az} for some a or, finally, of $\wp(z)$ and $\wp'(z)$. By Weierstrass all admit of algebraic addition theorems. When the genus of (5.2) is 0 there is a representation

= $R_1(s)$, $y = R_2(s)$ and hence a DE for $s(z)$

$$R_2'(s)\frac{ds}{dz} = R_2(s). \qquad (5.7)$$

nis must be a Riccati equation if (5.4) and hence also (5.7) have single-valued olutions. It follows that $s(z)$ and $w(z)$ are either rational in z or in e^{az}. f the genus is 1 we have instead $x = R_1(s, t)$, $y = R_2(s, t)$ with $^2 = 4s^3 - g_2 s - g_3$ and a DE for $s(z)$ which reduces to that of the $\wp$-function.

I have tried to imitate this procedure for $k = 2$ where one is led to second rder DE's for $s(z)$. Instead of invoking the Malmquist theorem, as with (5.7), ne must now fall back on the 50 odd second order equations which according to ainlevé and R. Gambier have fixed critical points. For the genus 0 the olutions appear to be of the same type as in the case $k = 1$. For the genus 1 here is still some doubt.

For $k = 1$ or 2 the Nevanlinna order of a meromorphic solution is ≤ 2 and 1 for entire solutions. The Painlevé determinateness theorem holds for B-equations with $k = 1$ and 2 even if there are infinitely multivalued olutions.

References

[1] Bank, S. Some results on analytic and meromorphic solutions of algebraic differential equations. Advances in Mathematics, 15 (1975) 41-62.

[2] Bank, S. and R. Kaufman, On meromorphic solutions of first-order differential equations. Comm. Math. Helvetici, 39 (51) (1976) 289-299.

[3] Briot, Ch. and J.-Cl. Bouquet, Intégration des équations différentielles au moyen des fonctions elliptiques. Journal École Polytechnique. 21 (36) (1856).

[4] Coppel, W. A. A survey of quadratic systems. J. Differential Equations. 2 (1966) 295-340.

[5] Emden, R. Gaskugeln, Anwendungen der mechanischen Wärmetheorie auf Kosmologie und meteorologischen Probleme. Teubner, Leipzig, 1907.

[6] Fermi, E. Un metodo statistico per la determinazione di alcune proprietà dell'atome. Atti Accad. Naz. Lincei, 6 (1927) 602-7.

[7] Gol'dberg, A. A. On single-valued solutions of first-order differential equations. Ukrain. Mat. Žurnal, 8: 3 (1956) 254-261. (Russian, NRL Translation 1224, 1970.)

[8] Hille, E. On a class of series expansions in the theory of Emden's equation. Proc. R. Soc. Edinburgh (A), 71: 8 (1972/73) 95-110.

[9] ----- A note on quadratic systems. Proc. R. Soc. Edinburgh (A), 72: 3 (1972/73) 17-37.

[10] ----- Finiteness of the order of meromorphic solutions of some non-linear ordinary differential equations. Proc. R. Soc. Edinburgh. 72 (1973/74) 331-336.

[11] ----- On some generalizations of the Malmquist theorem. Math. Scand. 39 (1977) 59-79.

[12] Laine, I. Admissible solutions of some generalized algebraic differential equations. Publ. Univ. Joensuu, B. 10 (1974) 6 pp.

[13] Malmquist, J. Sur les fonctions à un nombre fini de branches définies par les équations différentielles du premier ordre. Acta Math. 30 (1913) 297-334.

[14] Smith. R. Singularities of solutions of certain plane autonomous systems. Proc. R. Soc. Edinburgh, (A) 72: 26 (1973/74) 307-315.

[15] Thomas, L. H. The calculation of atomic fields. Proc. Cambridge Philosophica Soc. 13 (1927) 542-548.

[16] Wittich, H. Neuere Untersuchungen über eindeutige analytische Funktionen (Ergebnisse Math. 8). Springer-Verlag 1955.

[17] Yang, C.-C. On meromorphic solutions of generalized algebraic differential equations. Ann. Mat. Pura Appl. (IV) 91 (1972) 41-52.

[18] Yosida, K. A generalization of a Malmquist's theorem. Japan. Journal of Math. 9 (1933) 253-256.

[19] ----- On the characteristic function of a transcendental meromorphic solution of an algebraic differential equation of the first order and the first degree. Proc. Math. Phys. Soc. Japan. (2) 15 (1933) 337-338.

University of California, San Diego

ON DIFFERENTIAL RELATIONS AND ON FILIPPOV'S CONCEPT OF DIFFERENTIAL EQUATIONS

J. Kurzweil

We are interested in locally absolutely continuous solutions of differential relations of the form

(1) $\dot{x} \in F(t,x)$

under the assumption that

(2) $F : G \to K_n$, $G \subset R \times R^n$ and K_n is the system of non-empty convex compact subsets of R^n.

It is well known that existence of solutions and continuous dependence on initial conditions (and the right-hand side) for (1) can be established provided that

(3) G is open ,

(4) $F(t,.)$ is uppersemicontinuous for almost all t (the map $F(t,.)$ being defined by $F(t,.)(x) = F(t,x)$) ,

(5) $F(.,x)$ is measurable for every $x \in R^n$ (i.e. the set $\{t \mid F(t,x) \cap P \neq \emptyset\}$ is measurable for every closed set $P \subset R^n$) ,

(6) F fulfils a standard boundedness condition (to every $(s,y) \in G$ there exist such a neighbourhood U , $(s,y) \in U \subset G$, and a locally integrable function r that $\|z\| \leq r(t)$ for $z \in F(t,x)$, $(t,x) \in U$) .

Let m_k denote the k-dimensional Lebesgue measure. Let G be open and let $f : G \to R^n$ fulfil the usual conditions for the existence of solutions of

(7) $\dot{x} = f(t,x)$.

Then there exists such a subset $A \subset R$ that $m_1(R-A) = 0$ and that for any solution $u : I \to R^n$ of (7) and any $s \in I \cap A$ the derivative $\dot{u}(s)$ exists and $\dot{u}(s) = f(s,u(s))$. For the proof see [1]; the proof is "constructive", i.e. the set A can be defined from the properties of f .

This result can be extended to differential relations, if the concept of the derivative is replaced by that of the contingence derivative (if $u : I \to R^n$, $s \in I$, then the contingence derivative $Du(s)$ (of u at s) is the set of $z \in R^n$ that there exists a sequence $t_j \to s$ such that $z = \lim_{j\to\infty} (u(t_j)-u(s))/(t_j-s))$.

<u>Theorem 1</u>: Let (2)-(6) hold. Then there exists such a set $A \subset R$ that $m_1(R-A) = 0$ and that $\emptyset \neq Du(s) \subset F(s,u(s))$ for any solution $u : I \to R^n$ of (1) and any $s \in I \cap A$.

However, the proof obtained by the author is a pure existence proof with no information about A. It is evident that

(8) to every $\varepsilon > 0$ there exists such a set $A_\varepsilon \subset R$ that $m_1(R-A_\varepsilon) < \varepsilon$ and that the restriction $F|_{G \cap (A_\varepsilon \times R^n)}$ is uppersemicontinuous (with respect to the couple (t,x))

implies (4) and (5); the inverse implication is false. If Theorem 1 is modified so that (2), (3), (6) and (8) is assumed, then we can have a "constructive" proof (see [2]). The difficulties in the proof of the original Theorem 1 are due to the fact that assumptions (2) - - (6) may be fulfilled by rather irregular functions F. In examining (1), assumption (4) can be always replaced by the more restrictive assumption (8) without loss of generality - this follows from Theorem 2. Denote by $Sol(F)$ the set of solutions of (1). Put $K_n^o = K_n \cup \{\emptyset\}$.

<u>Theorem 2</u>: Let (2) and (4) hold. Then there exists such a function $H : G \to K_n^o$ fulfilling (8) that

(9) $$H(t,x) \subset F(t,x) \quad \text{for} \quad (t,x) \in G,$$

(10) $$Sol(H) = Sol(F).$$

Theorem 2 follows directly from a more general theorem on selections, for which the reader is referred to [3]. Of course, if the existence theorem holds for (1) then the values of H are nonempty sets.

Write $F \cong J$ if $F,J : G \to K_n^o$ and if there exists such a set $A \subset R$ that $m_1(R-A) = 0$ and that $F(t,x) = J(t,x)$ for $(t,x) \in G$, $t \in A$. If $F \cong J$, then obviously $Sol(F) = Sol(J)$. However, it can happen that F,J fulfil (2), (3), (6) and (8), $Sol(F) = Sol(J)$, but $F \not\supset J$. To see this, let $n=1$, let $W \subset R$ be closed, $R-W$ dense and let w be the characterisric function of W; put $F(t,x) = \langle 0, w(x) \rangle$, $J(t,x) = \{0\}$ for $t,x \in R$. As $R-W$ is dense, all solutions of $\dot{x} \in F(t,x)$ are constant functions and F is uppersemicontinuous, as W is closed.

Let V be a set of locally absolutely continuous functions $v : I_v \to R^n$ (I_v being the definition interval of v) such that $(t, v(t)) \in G$ for $t \in I_v$.

Theorem 3: Let G be open and let the derivatives of functions from V fulfil a suitable boundedness condition (for simplicity it may be assumed that $\|\dot{v}(t)\| \leqq \varrho(t)$ a.e. for $v \in V$, ϱ being locally integrable).

Then there exists a function $L_V : G \to K_n^o$ fulfilling (8) that

(11) $V \subset \mathrm{Sol}(L_V)$

(12) if $H : G \to K_n^o$ fulfils (4) and if $V \subset \mathrm{Sol}(H)$, then there exists such a set $A \subset R$ that $m_1(R-A) = 0$ and that $L_V(t,x) \subset H(t,x)$ for $(t,x) \in G$, $t \in A$ (we may write $L_V \subset H \mod \cong$).

L_V is obviously unique $\mod \cong$. The proof of Theorem 3 may be sketched as follows: A countable "dense" subset $\{v_i \mid i=1,2,\dots\} \subset V$ is chosen and a candidate J for L_V is defined by

$$J(t,x) = \bigcap_{\delta>0} \overline{\mathrm{conv}} \{\dot{v}_i(t) \mid \|v_i(t)-x\| < \delta , \ i=1,2,\dots\} .$$

Then it is proved that J fulfils (4), (5), (6), that $V \subset \mathrm{Sol}(J)$ and that J fulfils (12). By Theorem 2, J fulfils (8).

Putting $V = \mathrm{Sol}(F)$ in Theorem 3, we obtain the following

Corollary: Let $F : G \to K_n$ fulfil (3), (6) and (8). Then there exists such a $W : G \to K_n$ that $\mathrm{Sol}(W) = \mathrm{Sol}(F)$ and

(13) if H fulfils (8) and $\mathrm{Sol}(H) = \mathrm{Sol}(F)$, then $W \subset H \mod \cong$.

Thus, up to the equivalence $\cong$, W is the canonical element in the class of such functions fulfilling (8) that $\mathrm{Sol}(H) = \mathrm{Sol}(F)$.

A.F.Filippov developed in [4] a theory of the differential equation (7) provided that f is only measurable with respect to the couple (t,x) and fulfils a standard boundedness condition. His approach can be described in a more modern way as follows:

Let $Q \subset R^n$ be open and let $q : Q \to R^n$ be locally bounded. Let Sq denote such a function $M : Q \to K_n$ that

(14) M is uppersemicontinuous ,

(15) $q(x) \in M(x)$ a.e. in Q ,

(16) if a function $J : Q \to K_n$ fulfils (14), (15), then $M(x) \subset J(x)$ for $x \in Q$.

Obviously, M is defined uniquely by (14)-(16). In order to prove that M exists, one may define

$$M(x) = \bigcap_{\delta, N} \overline{\text{conv}} \left\{ q(y) \mid y \in Q-N, \|y-x\| < \delta \right\}$$

(the intersection is taken for all positive δ and for all subsets $N \subset Q$ such that $m_n(N) = 0$) and verify that $M(x) \neq \emptyset$ and that (14)-(16) is fulfilled. For an alternative proof see [5] .

If $f : G \to R^n$ and $f(t,.)$ is locally bounded for every t , we can define the function $F = Zf : G \to K_n$ by

$$(Zf)(t,x) = (Sf(t,.))(x)$$

and define that $u : I \to R^n$ is a solution of (7), if it is a solution of (1). It can be proved that $F = Zf$ fulfils (8) if f is measurable. If f is measurable and fulfils a boundedness condition, then $F = Zf$ fulfils (6) and (8) and a theory of equation (7) is obtained from the well-known results on (1). (Filippov did not dispose of a theory of (1) and e.g. he found solutions of (7) as limits of convergent subsequences of solutions of $\dot{x} = f_i(t,x)$, f_i being smooth functions converging to f .)

The relation of Filippov's theory of (7) to the theory of (1) is described in the following

Theorem 4: Let $F : G \to K_n$, G open. Then the following conditions are equivalent:

(17) there exists such a measurable function $f : G \to R^n$ that $f(t,.)$ is locally bounded for almost all t and $F = Zf$ mod $\cong$,

(18) F fulfils (8) and there exists such a set $A \subset R$ that $m_1(R-A) = 0$ and if $t \in A$, if the function $P : \{x \mid (t,x) \in G\} \to K_n$ is uppersemicontinuous and $P(x) \supset F(t,x)$ a.e. in $\{x \mid (t,x) \in G\}$, then $P(x) \supset$

$$\supset F(t,x) \text{ everywhere in } \{x \mid (t,x) \in G\},$$

(19) $$F(t,x) = \bigcap_{\delta, N} \overline{\mathrm{conv}} \bigcup \{F(t,y) \mid \|y-x\| < \delta,\ y \notin N\}$$

the intersection being taken for all positive δ and all sets $N \subset R^n$ with $m_n(N) = 0$) .

References

[1] G.Scorza-Dragoni: Una applicazione della quasi-continuità semiregolare delle funzioni misurabili rispetto a una e continue rispetto ad un'altra variabile. Atti.Acc.Naz.Lincei XII (1952), 1, 55-61pp.

[2] J.Jarník,J.Kurzweil: Extension of a Scorza-Dragoni Theorem to Differential Relations and Functional Differential Relations. Commentationes Mathematicae (issue in honour of Władysław Orlicz) - to appear.

[3] J.Jarník,J.Kurzweil: On Conditions on Right Hand Sides of Differential Relations. Čas.pěst.mat. - to appear.

[4] A.F.Filippov: Differential Equations with Discontinuous Right Hand Sides (in Russian). Matematičeskij Sbornik, 51 (1960), 99-128pp.

[5] I.Vrkoč:A Filippov's notion of generalized differential equations - to appear.

Author's address: J.Kurzweil
Matematický ústav ČSAV
Žitná 25
115 67 Praha 1
ČSSR

STURM-LIOUVILLE PROBLEMS WITH INDEFINITE WEIGHT FUNCTION AND OPERATORS IN SPACES WITH INDEFINITE METRIC

Heinz Langer
Sektion Mathematik
Technische Universität Dresden, GDR

§1. Introduction.

We consider the Sturm-Liouville problem

$$M[y] := -(py')' + qy = \lambda ry, \; p(0)y'(0)\sin\alpha - y(0)\cos\alpha = 0, \qquad (1$$

on an interval $[0,\ell)$, $0 < \ell \leq \infty$, of the real axis, $0 \leq \alpha < \pi$. The functions p, q, r satisfy the following conditions:

$$p \in AC_{loc}[0,\ell), \quad p(x) > 0 \quad (x \in [0,\ell)),$$

$$q \in L_{loc}[0,\ell), \quad q(x) \geq 0 \quad (x \in [0,\ell)),$$

$$r \in L_{loc}[0,\ell), \int_0^x |r(t)|dt > 0, \; \infty \geq \int_x^\ell |r(t)|dt > 0 \; (x \in [0,\ell)).$$

Moreover, we suppose that r changes sign, that is there are subsets Δ_+, Δ_- of positive Lebesgue measure such that $r(x) > 0$ (< 0) if $x \in \Delta$ (Δ_- resp.) a.e.

The conditions on p, q are the same as e.g. in [1], our conditions r is, however, stronger. It allows us to avoid technical complicati and additional hypotheses (e.g. (B) in [2] and $q \geq K|r|$ in the H_α - setting of §2, comp. [1], [3]).

The problem (1) can be written formally as $Sy = \lambda Ty$ where

$$Sy := -(py')' + qy, \; p(0)y'(0)\sin\alpha - y(0)\cos\alpha = 0$$

and T is multiplication by r. Assume for a moment that r is nonnegat and $0 < \alpha \leq \pi/2$. Then according to Å. Pleijel (see e.g. [4], [5]), (1) can be treated in the framework of T-theory or S-theory, that is in Hilbert spaces with scalar products given by

$$\int_0^\ell f\bar{g}r\,dx \text{ or } \int_0^\ell (pf'\overline{g'} + qf\bar{g})dx + \cot\alpha \; f(0)\overline{g(0)} \text{ resp.} \qquad (2$$

In either case a classification limit-point or limit-circle holds. Under the above assumptions about p, q, r and α the first scalar pro duct in (2) is always indefinite; it is connected in a natural way with the Krein space L_r^2 (see [6], [7]). The second scalar product in (2) is positive definite if $0 \leq \alpha \leq \alpha_0$ and indefinite if $\alpha_0 < \alpha < \pi$ for some $\alpha_0 \in [\pi/2,\pi)$; it is connected with a Hilbert space or a π_1-space (Pontrjagin space with index one, see [6]) resp., which will be denoted by H_α (see §2).

The problem (1) can be treated in either of these spaces, that is, with (1) can be associated operators symmetric with respect to the scalar products (2). Under additional hypotheses these operators will be even selfadjoint. In Å. Pleijel's investigations limit-point in S-theory is not equivalent to limit-point in T-theory. Correspondingly, in our situation the conditions for self-adjointness of the operators associated with (1) in H_α and L_r^2 are different. But in both cases it is important, that these operators are definitizable *), therefore they admit some kind of spectral decomposition ([8], [9], [6]).

In [7] the problem (1) is treated in L_r^2. If e.g. one of the following conditions is satisfied:

(a) $q(x) \geq K|r(x)|$, $x \in [0,\ell)$ a.e. with some constant K,

(b) $r(x)$ is of constant sign if $x \geq \ell'$, $0 < \ell' < \ell$,

and, moreover, the problem

$$M[y] = \lambda|r|y,\ p(0)y'(0)\sin\alpha - y(0)\cos\alpha = 0$$

is limit-point at ℓ, then a spectral function can be associated with (1). Expansion theorems in L_r^2 or some space with negative norm follow (see [7]).

Here we consider (1) in the Hilbert or π_1-space H_α. If

$$\int_0^\ell (p^{-1} + q)dx = \infty , \tag{3}$$

then with (1) a selfadjoint or π-selfadjoint operator in H_α is associated. Expansion theorems in H_α or some space with negative norm follow (see §2.4). As in [7] Weyl's coefficient, originally introduced in [1], plays an essential role in these considerations.

Finally, in §3 we sketch some results on the eigenvalue problem

$$dy' + \lambda y dM = 0 , \quad y'(0)\sin\alpha - y(0)\cos\alpha = 0. \tag{4}$$

---

*) An operator A in a space with indefinite scalar product $[\cdot,\cdot]$ is called definitizable (positizable in the terminology of [6]) if there exists a polynomial $\mathcal{P}$ such that $[\mathcal{P}(A)f,f] \geq 0$ for all $f \in D(\mathcal{P}(A))$.

Here M is a real function on $[0,\ell)$ which is of bounded variation on each compact subinterval of $[0,\ell)$. As above this problem can be treated either in L^2_M *) or in $H_{\alpha,M}$ (see §3). In both cases necessary and sufficient conditions for the selfadjointness of the correspondin operators can be given. The problem (1) can be reduced to (4) (see [10], [11]).

§2. The problem (1) in the space H_α.

1. The space H_α. By H' we denote the linear space of all absolutely continuous functions f on $[0,\ell)$ such that $p^{1/2}f' \in L^2$, $q^{1/2}f \in L^2$. If $\alpha \in (0,\pi)$, H'_α is H' endowed with the scalar product

$$(f,g)_\alpha := \int_0^\ell (pf'\overline{g'} + qf\overline{g})dx + \cot\alpha\, f(0)\overline{g(0)} \quad (f,g \in H'),$$

H'_0 is the linear space of all $f \in H'$, $f(0) = 0$, with the scalar product

$$(f,g)_0 := \int_0^\ell (pf'\overline{g'} + qf\overline{g})dx \quad (f,g \in H',\ f(0) = g(0) = 0).$$

Lemma 1. There exists an $\alpha_0 \in [\pi/2,\pi)$, such that H'_α is a Hilbert space if $\alpha \in [0,\alpha_0)$ and a π_1-space if $\alpha \in (\alpha_0,\pi)$. H'_{α_0} is either a Hilbert space or the scalar product $(\cdot,\cdot)_{\alpha_0}$ degenerates on H'_{α_0} with a one-dimensional isotropic subspace $\mathcal{L}_0$; then the factor space $H'_{\alpha_0}/\mathcal{L}_0$ is a Hilbert space.

For $\alpha \neq \alpha_0$ the convergence of a sequence (f_n) to zero in the norm of H'_α is equivalent to

$$f_n(0) \to 0,\ \int_0^\ell (p|f_n'|^2 + q|f_n|^2)dx \to 0.$$

Define (comp. [3])

$$G := \{g \in H' :\ r(x)g(x) = 0 \ \text{ a.e. on } [0,\ell)\}.$$

The conditions on r imply $g(0) = 0$ for all $g \in G$. Evidently $(g,g)_\alpha \geqq 0$, and if the equality sign holds we have $g'(x) = 0$ a.e., therefore $g(x) = 0$ $(x \in [0,\ell))$. Moreover, G is closed in H'_α if $(\cdot,\cdot)_\alpha$ is nondegenerated and the image of G in $H'_{\alpha_0}/\mathcal{L}_0$ is closed if $(\cdot,\cdot)_{\alpha_0}$ degenerates.

*) The notation L^2_M is not consistent with L^2_r, but we hope no confusion will arise.

In the following we shall consider the spaces H_α:

$H_\alpha := \{f \in H'_\alpha : (f,G)_\alpha = \{0\}\}$

if $(\cdot,\cdot)_\alpha$ is nondegenerated and

$H_{\alpha_0} := \{f \in H'_{\alpha_0}/\mathcal{L}_0 : (f,G)_{\alpha_0} = \{0\}\}$

otherwise. The properties of G imply

Lemma 2. H_α is a Hilbert space if $\alpha \in [0,\alpha_0]$ and a π_1-space if $\alpha \in (\alpha_0,\pi)$.

The set H_α, $\alpha > 0$, or the union of the cosets of H_{α_0} is independent of α because we have $(f,g)_\alpha = (f,g)_{\pi/2}$ ($f \in H'$, $g \in G$).

2. The operator A_α. In H_α we consider the following operator A_α: $D(A_\alpha)$ is the set of all $f \in H_\alpha$ such that

(i) pf' is absolutely continuous on $[0,\ell)$ and $M[f] = rg$ a.e. with some $g \in H_\alpha$; (5)

(ii) $(pf')(0)\sin\alpha - f(0)\cos\alpha = 0$;

(iii) f vanishes at ℓ *)

and $A_\alpha f := g$ if (5) holds, $f \in D(A_\alpha)$.

If $(\cdot,\cdot)_\alpha$ degenerates, it is not hard to see that the functions v_0: $(v_0,v_0^0)_{\alpha_0} = 0$ satisfy $M[v_0] = 0$ and the boundary condition (ii). In this case $D(A_{\alpha_0})$ is, more exactly, the set of all cosets $f \in H_{\alpha_0}$ which contain at least one element f with properties (i - iii).

This definition is correct . Indeed, $M[f] = rg_1$ and $M[f] = rg_2$, $g_1, g_2 \in H_\alpha$, imply $r(g_1-g_2) = 0$ a.e. or $g_1-g_2 \in G$, that is $g_1 = g_2$.

Theorem 3. Suppose (3) holds. Then $D(A_\alpha)$ is dense in H_α and the closure $\overline{A_\alpha}$ of A_α is selfadjoint (π-selfadjoint) in H_α if $\alpha \in [0,\alpha_0]$ ($\alpha \in (\alpha_0,\pi)$ resp.).

3. The resolvent of $\overline{A_\alpha}$; Weyl's coefficient. In sections 2.3–4 we always suppose (3). Then, if $\lambda \neq \overline{\lambda}$, $\lambda \notin \sigma_p(\overline{A_\alpha})$, the equation $(\overline{A_\alpha} - \lambda I)f = g$ for arbitrary $g \in H_\alpha$ has exactly one solution

---

*) That is there exists an $\ell' < \ell$, depending of f, such that $f(x) = 0$ if $\ell' \leqq x < \ell$.

$f =: R_\lambda^{(\alpha)} g = R_\lambda g$ *) $\in H_\alpha$. It is the solution in H_α of the initial problem

$$M[f] - \lambda r f = rg,\ p(0)f'(0)\sin\alpha - f(0)\cos\alpha = 0.$$

By φ_α, θ_α we denote the solutions of the equation $M[y] = \lambda r y$ satisfying the conditions

$$\varphi_\alpha(0;\lambda) = \sin\alpha \ ,\quad p(0)\ \varphi_\alpha'(0;\lambda) = \cos\alpha,$$

$$\theta_\alpha(0;\lambda) = \cos\alpha \ ,\quad p(0)\ \theta_\alpha'(0;\lambda) = -\sin\alpha.$$

Considering $g \in H_\alpha$ vanishing at ℓ and such that $\int_0^\ell g\varphi_\alpha r dx \neq 0$, it is not hard to see (comp. [7]) that there exists a function m_α, piecewise meromorphic in the upper and lower half planes, such that the function $\psi_\alpha(\cdot;\lambda)$:

$$\psi_\alpha(x;\lambda) := \theta_\alpha(x;\lambda) + m_\alpha(\lambda)\ \varphi_\alpha(x;\lambda)$$

is (up to scalar multiples) the unique solution in H_α of $M[y] = \lambda r y$ ($\operatorname{Im}\lambda \neq 0$, $\lambda \notin \sigma_p(\overline{A_\alpha})$). Now it is easy to check that for $g \in H_\alpha$ vanishing at ℓ:

$$(R_\lambda g)(x) = \psi_\alpha(x;\lambda)\int_0^x g\ \varphi_\alpha r\ d\xi + \varphi_\alpha(x;\lambda)\int_x^\ell g\ \psi_\alpha r d\xi,$$

$$(R_\lambda g,g)_\alpha = \lambda m_\alpha(\lambda)\ \Phi_\alpha(g;\lambda)\ \Phi_\alpha(\overline{g};\lambda) + \dots \ , \qquad (6)$$

where $\Phi_\alpha(g;\lambda) := \int_0^\ell \varphi_\alpha\ grdx$ is the Fourier transform of g and ... denotes an entire function.

A complex function F belongs to class $N_\varkappa$ ($\varkappa$ - nonnegative integer) if it has the following properties (see [12], [7]):

(i) It is defined and meromorphic in the upper half plane C_+ ;

(ii) the kernel $K_F(z,\zeta) := (F(z)-\overline{F(\zeta)})(z-\overline{\zeta})^{-1}$ ($z,\zeta \in C_+$) has exactly $\varkappa$ negative squares.

We always extend $F \in N_\varkappa$ to the lower half plane by $F(\overline{z}) = \overline{F(z)}$ ($z \in C_+$).

Define $\hat{m}_\alpha$:

$$\hat{m}_\alpha(\lambda) := \begin{cases} \lambda(m_\alpha(\lambda) + \cot\alpha) & \alpha \neq 0\ , \\ \lambda^{-1} m_0(\lambda) & \alpha = 0\ . \end{cases}$$

<u>Theorem 4</u>. If $\alpha \in [0,\alpha_0]$ ($\alpha \in (\alpha_0,\pi)$) we have $\hat{m}_\alpha \in N_0$ ($\hat{m}_\alpha \in N_1$ resp.). If $\hat{m}_\alpha \in N_1$ there exists a complex number $s_\alpha \neq 0$ such that one and only one of the following representations holds (see [12], [7]):

---

*) The upper index α will often be omitted in the following.

$$1)\ s_\alpha \neq \overline{s_\alpha}\ :\ \hat{m}_\alpha(\lambda) = c_0 + c_1\lambda + \int_{-\infty}^{\infty}((t-\lambda)^{-1} - t(1+t^2)^{-1}\chi_\infty(t))(1+t^2)d\sigma_\alpha(t) + a(\lambda-s_\alpha)^{-1} + \bar{a}(\lambda-\overline{s_\alpha})^{-1}\ ;$$

$2)\ s_\alpha = \overline{s_\alpha}$:

$$2a)\ \hat{m}_\alpha(\lambda)=c_0+c_1\lambda+\int_{-\infty}^{\infty}((t-\lambda)^{-1}-t(1+t^2)^{-1}\chi_\infty(t))(1+t^2)d\sigma_\alpha(t) + b_1(s_\alpha-\lambda)^{-1}\ ;$$

$$2b)\ \hat{m}_\alpha(\lambda)=c_0+c_1\lambda+\int_{-\infty}^{\infty}((t-\lambda)^{-1}-t(1+t^2)^{-1}\chi_\infty(t))(1+t^2)d\sigma_\alpha(t) + d_1(s_\alpha-\lambda)^{-1}+d_2(s_\alpha-\lambda)^{-2}+d_3(s_\alpha-\lambda)^{-3}\ ;$$

$$2c)\ \hat{m}_\alpha(\lambda)=c_0+c_1\lambda+\int_{-\infty}^{\infty}((t-\lambda)^{-1}-(t+\lambda)(1+t^2)^{-1}\chi_\infty(t) + \sum_1^2(t-s_\alpha)^{j-1}(\lambda-s_\alpha)^{-j}\chi_\alpha(t))\cdot$$
$$\cdot(1+t^2)^2(t-s_\alpha)^{-2}d\sigma_\alpha(t)+d_1(s_\alpha-\lambda)^{-1}+d_2(s_\alpha-\lambda)^{-2}+d_3(s_\alpha-\lambda)^{-3}. \qquad (7)$$

Here σ_α is a left continuous nondecreasing bounded function, continuous in s_α if $s_\alpha = \overline{s_\alpha}$ and $\int\chi_\alpha(t)(t-s_\alpha)^{-2}d\sigma_\alpha(t) = \infty$ in case 2c); $c_1,\ d_3 \geqq 0$, $b_1 < 0$; $c_0,\ d_1,\ d_2$ are real, a is complex and $|d_2|+|d_3| \neq 0$ in case 2b); χ_α (χ_∞) is the characteristic function of a neighbourhood Δ_α (Δ_∞) of s_α (∞ resp.), $\Delta_\alpha \cap \Delta_\infty = \emptyset$, $0 \notin \overline{\Delta_\alpha}$.

4. Expansion theorems. If $\alpha \in [0,\alpha_0]$ the operator $\overline{A_\alpha}$ is selfadjoint in the Hilbert space H_α and expansion theorems follow in the usual way. Therefore we suppose in this section $\alpha \in (\alpha_0,\pi)$. Then $\overline{A_\alpha}$ is π-selfadjoint in the π_1-space H_α. A theorem by L.S. Pontrjagin [6] implies that $\overline{A_\alpha}$ has one and only one eigenvalue s_α with eigenelement f_α, such that $(f_\alpha,f_\alpha)_\alpha \leqq 0$. It can be shown that s_α coincides with the number s_α in the representation of $\hat{m}_\alpha$.

If f, g $\in H_\alpha$ vanishing at ℓ we have from (6)

$$(R_\lambda f,g)_\alpha = \hat{m}_\alpha(\lambda)\Phi_\alpha(f;\lambda)\Phi_\alpha(\bar{g};\lambda) + \ldots \qquad (8)$$

and if f, g $\in D(A_\alpha)$

$$(R_\lambda(A_\alpha-s_\alpha I)f,\ (A_\alpha-s_\alpha I)g)_\alpha = (\lambda-s_\alpha)(\lambda-\overline{s_\alpha})\hat{m}_\alpha(\lambda)\Phi_\alpha(f;\lambda)\Phi_\alpha(\bar{g};\lambda) + \ldots\ , \qquad (9)$$

where ... again denotes an entire function.

Suppose first $s_\alpha \neq \overline{s_\alpha}$ or that s_α is a (real) regular critical point of $\overline{A_\alpha}$ (see [9], [7]), P_α denotes the π-orthogonal projector onto the algebraic eigenspace of $\overline{A_\alpha}$ corresponding to s_α. Then $\hat{m}_\alpha$ admits one of the representations 1), 2a), 2b) above, and (8) and the Stieltjes-Livšic inversion formula imply for an arbitrary open interval Δ and $f, g \in H_\alpha$ vanishing at ℓ:

$$(E_\alpha(\Delta)f,g)_\alpha = \int_\Delta (1+t^2)\Phi_\alpha(f;t)\Phi_\alpha(\bar{g};t)d\sigma_\alpha(t) \text{ in case 1),} \quad (10)$$

$$(E_\alpha(\Delta)f,g)_\alpha - (P_\alpha f,g)_\alpha \chi_\Delta(s_\alpha) = \int_\Delta (1+t^2)\Phi_\alpha(f;t)\Phi_\alpha(\bar{g};t)d\sigma_\alpha(t) \text{ in cases 2a), 2b);} \quad (11)$$

here E_α denotes the spectral function of $\bar{A}_\alpha$. Hence the Fourier transformation can be extended to all of H_α, the images $\Phi_\alpha(f;\cdot)$, $f \in H_\alpha$, being in $L^2_{\tilde{\sigma}_\alpha}$, where $d\tilde{\sigma}_\alpha(t) = (1+t^2)d\sigma_\alpha(t)$, and (10), (11) hold true for all $f, g \in H_\alpha$.

The function $\varphi_\alpha(\cdot;s_\alpha)$ is the eigenvector of $\overline{A_\alpha}$ corresponding to the eigenvalue s_α; we denote it by φ_α and $\varphi_\alpha(\cdot;\overline{s_\alpha})$ by $\overline{\varphi_\alpha}$. Moreover, if $s_\alpha = \overline{s_\alpha}$ there may be a Jordan chain of length at most 3, its element being $\dot{\varphi}_\alpha$ $\ddot{\varphi}_\alpha$, where e.g. $\dot{\varphi}_\alpha(x) = \left.\frac{\partial\varphi_\alpha(x;\lambda)}{\partial\lambda}\right|_{\lambda=s_\alpha}$.

Theorem 5. 1) Suppose the π-selfadjoint operator $\overline{A_\alpha}$ has a pair of nonreal eigenvalues s_α, $\overline{s_\alpha}$. Then for an arbitrary Borel set $\Delta \subset R^1$ and $f \in H_\alpha$ we have

$$(E_\alpha(\Delta)f)(x) = \int_\Delta \Phi_\alpha(f;t)\varphi_\alpha(x;t)t^{-1}d\tilde{\sigma}_\alpha(t),$$

$$f(x) = \int_{R^1} \Phi_\alpha(f;t)\varphi_\alpha(x;t)t^{-1}d\tilde{\sigma}_\alpha(t)+(f,\overline{\varphi_\alpha})_\alpha(\varphi_\alpha,\overline{\varphi_\alpha})_\alpha^{-1}\varphi_\alpha(x) +(f,\varphi_\alpha)_\alpha(\overline{\varphi_\alpha},\varphi_\alpha)_\alpha^{-1}\,\overline{\varphi_\alpha}(x).$$

2) Suppose $s_\alpha = \overline{s_\alpha}$ and s_α is a regular critical point of $\overline{A}_\alpha$. Then fo[r] an arbitrary Borel set $\Delta \subset R^1$ and $f \in H_\alpha$ we have

2a) $(E_\alpha(\Delta)f)(x) = \int_\Delta \Phi_\alpha(f;t)\varphi_\alpha(x;t)t^{-1}d\tilde{\sigma}_\alpha(t) +$

$+ \chi_\Delta(s_\alpha)(f,\varphi_\alpha)_\alpha(\varphi_\alpha,\varphi_\alpha)_\alpha\ \varphi_\alpha(x)$ if s_α is an algebraically simpl[e] eigenvalue of $\overline{A_\alpha}$; here $(\varphi_\alpha,\varphi_\alpha)_\alpha < 0$ and $E_\alpha(R^1)f = f$;

2b) if s_α is of algebraic multiplicity 2 (or 3), the functions $\dot{\varphi}_\alpha$ (and $\ddot{\varphi}_\alpha$ resp.) belong to H_α and with bounded linear functional[s] d_1, d_2 (and d_3 resp.) the representation

$(E_\alpha(\Delta)f)(x) = \int_\Delta \Phi_\alpha(f;t)\varphi_\alpha(x;t)t^{-1}d\tilde{\sigma}_\alpha(t) +$

$+ \chi_\Delta(s_\alpha)\{d_1(f)\varphi_\alpha(x)+d_2(f)\dot{\varphi}_\alpha(x)+d_3(f)\ddot{\varphi}_\alpha(x)\}$

holds; here $(\varphi_\alpha,\varphi_\alpha)_\alpha = 0$, $E_\alpha(R^1)f = f$.

If $0 \notin \bar{\Delta}$ and Δ is bounded the integrals exist for each $x \in [0,\ell)$; at 0 and ∞ they exist as improper integrals in the norm of H_α.

Remark 6. If the spectrum of $\overline{A_\alpha}$ is discrete in the finite complex plane, s_α cannot be a singular critical point, that is one of the cases 1), 2a), 2b) of Theorem 5 holds. Then the integrals simplify to series converging in the norm of H_α.

Remark 7. As in [7] it can be shown that for given coefficients p, q, r the case 1): $s_\alpha \neq \overline{s_\alpha}$ holds for infinitely many α's which form an open subset of (α_0,π).

Let now $s_\alpha = \overline{s_\alpha}$ be a singular critical point of $\overline{A_\alpha}$, that is $\hat{m}_\alpha$ admits the representation 2c). The bounded neighbourhood Δ_α of s_α, $0 \notin \overline{\Delta_\alpha}$, is the same as in (7). Let U be an open set in the complex plane, $U \supset \Delta_\alpha$, and define

$$\varrho(x) := \sup_{t\in U} \max\{p(x)|\varphi_\alpha'(x;t)|^2 + q(x)|\varphi_\alpha(x;t)|^2,$$
$$p(x)|\dot{\varphi}_\alpha'(x;t)|^2+q(x)|\dot{\varphi}_\alpha(x,t)|^2, p(x)|\ddot{\varphi}_\alpha'(x;t)|^2+q(x)|\ddot{\varphi}_\alpha(x,t)|^2\}.$$

With the function $\nu : \nu(x) := (1+x^2)(1+\varrho(x)^2)$ we introduce the space $H_{\alpha,+}$ of all $f \in H_\alpha$ such that

$$\|f\|_+^2 := \int_0^\ell (p|f'|^2 + q|f|^2)\nu\, dx + \cot\alpha \mid f(0)|^2 < \infty .$$

Then we have for $g \in H_\alpha$:

$$\|g\|_-^2 := \sup_{f\in H_+, f\neq 0} |(f,g)_\alpha|^2 \|f\|_+^{-2} \leq \int_0^\ell (p|g'|^2+q|g|^2)\nu^{-1}dx + |\cot\alpha|\,|g(0)|^2,$$

the functions $\varphi_\alpha(\cdot;t)$, $\dot{\varphi}_\alpha(\cdot;t)$, $\ddot{\varphi}_\alpha(\cdot;t)$ belong to the space $H_{\alpha,-}$ if $t \in U$, and $\|\varphi_\alpha(\cdot;t)\|_-$, $\|\varphi_\alpha(\cdot;t)\|_-$ and $\|\ddot{\varphi}_\alpha(\cdot;t)\|_-$ are uniformly bounded for all $t \in U$.

If f, $g \in D(A_\alpha)$, (9), (7) and the inversion formula imply

$$(E_\alpha(\Delta_\alpha)(\overline{A_\alpha}-s_\alpha I)f, (\overline{A_\alpha}-s_\alpha I)g)_\alpha =$$
$$= \int_{\Delta_\alpha} \Phi_\alpha(f;t)\Phi_\alpha(\bar{g};t)(1+t^2)^2\, d\sigma_\alpha(t) + d_3\Phi_\alpha(f;s_\alpha)\Phi_\alpha(\bar{g};s_\alpha).$$

In the same way we get from (8) and (7) for arbitrary f, $g \in H_\alpha$ vanishing at ℓ

$$(E_\alpha(R^1\setminus \Delta_\alpha)f,g)_\alpha = \int_{R^1\setminus\Delta_\alpha} \Phi_\alpha(f;t)\, \Phi_\alpha(\bar{g};t)(1+t^2)^2(t-s_\alpha)^{-2}d\sigma_\alpha(t).$$

The left hand sides of these relations evidently extend to H_α by continuity, hence the Fourier transformation can also be extended such that its images belong to $L^2_{\hat{\sigma}_\alpha}$, where

$$d\hat{\sigma}_\alpha(t) := d\tilde{\sigma}_\alpha(t)+d_3 d\delta_{s_\alpha}(t),\quad d\tilde{\sigma}_\alpha(t) := \begin{cases}(1+t^2)^2 d\sigma_\alpha(t) & t \in \Delta_\alpha \\ (1+t^2)^2(t-s_\alpha)^{-2}d\sigma_\alpha(t) & t \in R^1\setminus\Delta_\alpha\end{cases}$$

Theorem 8. Suppose $s_\alpha \neq \overline{s_\alpha}$ is a singular critical point of $\overline{A_\alpha}$. Then for an arbitrary Borel set $\Delta \subset R^1$ and $f \in H_\alpha$ we have

$$\begin{aligned}(E_\alpha(\Delta)f)(x) = \int_{\Delta\cap\Delta_\alpha} & \Phi_\alpha(f;t)\{\varphi_\alpha(x;t)-\varphi_\alpha(x;s_\alpha)-(t-s_\alpha)\dot{\varphi}_\alpha(t;s_\alpha)\} \\ & \cdot\ t^{-1}(t-s_\alpha)^{-2}d\tilde{\sigma}_\alpha(t) \\ & + \chi_\Delta(s_\alpha)\{d_1(f)\varphi_\alpha(x) + d_2(f)\dot{\varphi}_\alpha(x) + d_3(f)\ddot{\varphi}_\alpha(x)\} \\ & + \int_{\Delta\setminus\Delta_\alpha} \Phi_\alpha(f;t)\varphi_\alpha(x;t)t^{-1}d\tilde{\sigma}_\alpha(t). \qquad (12)\end{aligned}$$

If $0 \notin \overline{\Delta}$ and Δ is bounded the integrals exist for each $x \in [0,\ell)$, at 0 and ∞ the second integral on the right hand side exists as an improper integral in the norm of H_α. The first integral an the right hand side of (12) defines an element of $H_{\alpha,-}$, d_1, d_2, d_3 are continuous functionals on H_α.

§3. Generalized second order differential operators with indefinite weight function.

1. The problem (4) in L^2_M. Let M be a real-valued function on $[0,\ell)$ which is of bounded total variation $m(x)$ on each compact subinterval $[0,x]$ of $[0,\ell)$. Suppose that M is continuous at 0 but not constant a 0 and ℓ. In L^2_m we define the following scalar products:

$$(f,g) := \int_0^\ell f\bar{g}\,dm\,,\quad [f,g] := \int_0^\ell f\bar{g}dM \quad (f,g \in L^2_m).$$

They turn L^2_m into a Krein space denoted by L^2_M. The generalized secon order differential operator $D_M D_x$ in L^2_M is defined as in [13], [11] (see also [10]). By B_α, $0 \leq \alpha < \pi$, we denote the following operator in L^2_M:

$$D(B_\alpha) := \{f \in D(D_M D_x): f'(0)\sin\alpha - f(0)\cos\alpha = 0\},$$
$$B_\alpha f := -D_M D_x f \quad (f \in D(B_\alpha))\,.$$

Then B_α is J-symmetric, that is symmetric with respect to the scalar product $[\cdot,\cdot]$. It is J-selfadjoint if and only if $\int_0^\ell x^2 dm(x) = \infty$

In [13] it was shown that B_α is definitizable. If we assure by an additional hypothesis that the resolvent set of B_α is nonempty, expansion theorems in L^2_M for the differential problem (4) can be proved by the method used in [7].

2. The problem (4) in $H_{\alpha,M}$. By $\tilde{H}_{\alpha,M}$ we denote the space of all complex functions f on $[0,\ell)$ with the properties:

(i) f is absolutely continuous on $[0,\ell)$,
$f' \in L^2$ and $f(0) = 0$ if $\alpha = 0$;

(ii) f is linear on each interval where M is constant,

and with the scalar product

$$(f,g)_\alpha := \int_0^\ell f'\overline{g'}dx + \cot\alpha\, f(0)\overline{g(0)}\ , \qquad \alpha \in (0,\pi)\ ,$$

$$(f,g)_0 := \int_0^\ell f'\overline{g'}dx\ .$$

Consider the function v_α: $v_\alpha(x) = x\cos\alpha + \sin\alpha$. We have $v_\alpha \in \tilde{H}_{\alpha,M}$ if and only if $\ell < \infty$. In this case, if $\tan\alpha \neq -\ell$ we denote by $H_{\alpha,M}$ the space of all $f \in \tilde{H}_{\alpha,M}$ such that $f(\ell) = \lim_{x\uparrow\ell} f(x) = 0$; if $\tan = -\ell$: $H_{\alpha,M}$ is the factorspace of these $f \in H_{\alpha,M}$ modulo v_α. Finally if $\ell = \infty$ we put $H_{\alpha,M} = \tilde{H}_{\alpha,M}$. Then there exists an $\alpha_0 \in [\pi/2,\pi)$ such that $H_{\alpha,M}$ is a Hilbert space (π_1-space) if $\alpha \in [0,\alpha_0]$ ($\alpha \in (\alpha_0,\pi)$ resp.). In $H_{\alpha,M}$ we define an operator A_α in the following way:

$D(A_\alpha)$ is the set of all $f \in H_{\alpha,M}$ of the form $f(x) = \int_x^\ell (x-s)\varphi(s)dM(s)$ with some $\varphi \in H_{\alpha,M}$ vanishing at ℓ, $f'(0)\sin\alpha - f(0)\cos\alpha = 0$. In this case $A_\alpha f := \varphi$.

It is not hard to see that A_α is a densely defined symmetric or π-symmetric operator in $H_{\alpha,M}$.

<u>Theorem 9</u> (see [14]). The closure $\overline{A_\alpha}$ of A_α is selfadjoint if and only if

$$\ell + \int_0^\ell M(x)^2 dx = \infty\ .$$

In this case expansion theorems in $H_{\alpha,M}$ for the differential problem (4) follow in the same way as in §2 above.

Literature:

[1] F.V. Atkinson, W.N. Everitt, K.S. Ong. On the m-coefficient of Weyl for a differential equation with an indefinite weight function. Proc. Lond. Math. Soc. 29 (1974), 368-384.

[2] K. Daho, H. Langer. Some remarks on a paper by W.N. Everitt. Proc. Royal Soc. Edinburgh (A) (to appear).

[3] W.N. Everitt. Some remarks on a differential expression with an indefinite weight function. North Holland Math. Studies 13 (1974), 13-28.

[4] Å. Pleijel. Generalized Weyl circles. Lecture Notes in Math. 415 (1974), 211-226.

[5] Å. Pleijel. Generalization of Weyl's method to pairs of symmetric ordinary differential operators.

[6] J. Bognár. Indefinite inner product spaces (Berlin-Heidelberg-New York: Springer 1974).

[7] K. Daho, H. Langer. Sturm-Liouville operators with an indefinite weight function. Submitted to Proc. Royal Soc. Edinburgh (A).

[8] H. Langer. Spektraltheorie linearer Operatoren in J-Räumen und einige Anwendungen auf die Schar $L(\lambda) = \lambda^2 I + \lambda B + C$. Technische Universität Dresden, Habilitationsschrift 1965.

[9] M.G. Krein, H. Langer. On the spectral function of a self-adjoint operator in a space with indefinite matric. Dokl. Akad. Nauk SSSR 152, 39-42 (1963) [Russisch] (English transl. Soviet Math. Dokl. 4 (1963), 1236-1239).

[10] I.S. Kac, M.G. Krein. On the spectral functions of the string. Am. Math. Soc. Transl. (2) 103 (1974), 19-102.

[11] H. Langer. Spektralfunktionen einer Klasse von Differentialoperatoren zweiter Ordning mit nichtlinearem Eigenwertparameter. Ann. Acad. Sci. Fenn. Ser. AI, 2 (1976), 269-301.

[12] M.G. Krein, H. Langer. Über einige Fortsetzungsprobleme, die eng mit der Theorie hermitescher Operatoren im Raume Π_n zusammenhängen. Teil I: Einige Funktionenklassen und ihre Darstellungen. Math. Nachr. 77 (1977), 187-236.

[13] H. Langer. Zur Spektraltheorie verallgemeinerter gewöhnlicher Differentialoperatoren zweiter Ordnung mit einer nichtmonotonen Gewichtsfunktion. - Universität Jyväskylä, Mathematisches Institut, Bericht 14, Jyväskylä, 1972, 1-58.

[14] M.G. Krein, H. Langer, to be submitted to Proc. Royal Soc. Edinburgh (A).

Elementary comparison techniques for certain classes of Sturm-Liouville equations

Lee Lorch

1. Some background.

Many important functions satisfy Sturm-Liouville differential equations $(gy')' + f_\nu(x)y = 0$, $g(x) > 0$, where $g(x)$ and $f_\nu(x)$ are continuous, $a < x < b$, with $f_\nu(x)$ monotonic in both x and ν. Among them are Bessel functions, Coulomb wave functions, confluent hypergeometric functions and the classical orthogonal polynomials.

This provides an opportunity to infer much useful information about such functions from the Sturm theory [8] and its numerous extensions and analogues, many of which are recorded in [9]. Systematic use has been made of monotonicity in either x or ν. There is also an interesting application utilizing simultaneously both monotonicities: G. Szegö and P. Turán [11] showed that the differences of successive positive zeros of the ultraspherical polynomials $P_n^{(\lambda)}(x)$ are monotonic in the degree n, $0 < \lambda < 1$.

Sturm himself had used monotonicity in the parameter to show that a Bessel function has infinitely many zeros, and in x to show that the differences of the positive zeros, say $c_{\nu k}$, $k = 1,2,\ldots$, of an arbitrary Bessel function of fixed order ν, $C_\nu(x)$, are monotonic in k: $\Delta^2 c_{\nu k} > 0$, if $|\nu| < \frac{1}{2}$, $\Delta^2 c_{\nu k} < 0$, if $|\nu| > \frac{1}{2}$.

G.N. Watson [12, pp. 518-521] used a sharpened form of the Sturm comparison theorem to show that the first positive zero, $j_{\nu 1}$, of the Bessel function $J_\nu(x)$ is $\nu + k_1\nu^{1/3} + \underline{O}(\nu^{-1/3})$, as $\nu \to \infty$, an application in the parameter.

A similar form of the same theorem, but extended so as to permit a singular endpoint, was reached by E. Makai [6] who used it to obtain elegant proofs of the monotonicity of the areas under graphs of solutions of appropriate equations. This yielded a proof of a conjecture in quantum mechanics that, for the linear harmonic oscillator, the probability of finding the oscillating particle between two consecutive nodes is greater the further these nodes are from the origin.

These various approaches will be illustrated below by some commentary and by additional elementary applications, in some instances providing new results, in others, alternative proofs (perhaps simpler), intended to suggest the continuing value of the Sturm approach in the study of important special functions.

Sturm methods have been applied by so many different workers that it is not practicable to cite many names. The bibliography in [9] should be consulted. Szegö's book [10] lists many applications to the zeros of orthogonal polynomials.

2. The Sturmian formulations used here.

First, define the symbol " $\lessdot$ " as follows:

$f(x) \lessdot F(x)$, $a < x < b$, means $f(x) \leqq F(x)$, $a < x < b$;
and, for every $\varepsilon > 0$, $f(x) \not\equiv F(x)$, $a < x < a+\varepsilon$.

One applicable formulation of a Sturm theorem, incorporating an end-point condition introduced by G. Szegö [cf. 10, §1.82 and references] which extends the scope of such theorems to singular equations, reads:

Theorem 2.1. Let $f(x)$, $F(x)$, $g(x) > 0$, be continuous and $f(x) \lessdot F(x)$, $a < x < b$. Let $y(x)$, $Y(x)$ be positive solutions of, respectively,

$$(gy')' + f(x)y = 0, \quad (gY')' + F(x)Y = 0, \quad a < x < b,$$

and suppose that

$$\lim_{x \to a+} g(x)\{y'(x)Y(x) - y(x)Y'(x)\} \text{ exists and } \geqq 0 . \tag{2.1}$$

Then

$$\frac{y'(x)}{y(x)} > \frac{Y'(x)}{Y(x)}, \quad a < x < b . \tag{2.2}$$

The proof is an immediate consequence of Green's formula [2, p. 225] applied to $a < x_1 < x < b$, then letting $x_1 \to a+$.

A generalization of the theorem in which gY' is replaced by GY', with, say, $g(x) \geqq G(x)$, is an equally immediate consequence of Picone's formula [2, p. 226].

Another version to be used below, essentially Szegö's original formulation [10, Theorem 1.82.1, p. 19], states:

Theorem 2.2. With $f(x)$, $F(x)$ and $g(x)$ as in Theorem 2.1, let $y(x)$ and $Y(x)$ be arbitrary nontrivial solutions of the respective differential equations, with a and b consecutive zeros of $y(x)$. Then $Y(x)$ vanishes at least once in $a < x < b$.

3. On the Szegö end condition (2.1); some monotonicity results in the parameter.

Theorem 2.2 yields a simple proof (essentially going back to M. Bôcher and M.B. Porter [1]) of the familiar result

$$j_{\nu 1} < j_{\nu+\varepsilon,1} \quad \text{for } \nu \geqq 0 \text{ and } \varepsilon > 0 , \tag{3.1}$$

which can be extended to $j_{\nu k}$, $k = 2,3,\ldots$, via the interlacing properties of these eros. Let $g(x) = 1$, $f(x) = 1 - x^{-2}[(\nu+\varepsilon)^2 - \frac{1}{4}]$, $F(x) = 1 - x^{-2}(\nu^2 - \frac{1}{4})$, $x) = x^{1/2}J_{\nu+\varepsilon}(x)$, $Y(x) = x^{1/2}J_\nu(x)$, $a = 0$, $b = j_{\nu 1}$. Theorem 2.2 now gives (3.1), nce $J_\nu(x) \sim x^\nu$ as $x \to 0+$.

Theorem 2.1 implies

$$\text{(3.2)} \qquad \frac{J'_\nu(x)}{J_\nu(x)} < \frac{J'_{\nu+\varepsilon}(x)}{J_{\nu+\varepsilon}(x)}, \quad 0 < x < j_{\nu 1}, \ \nu \geqq 0, \quad \varepsilon > 0,$$

hich can be extended, for $c_{\nu+\varepsilon,k-1} < c_{\nu k}$ and $C_\nu(x)$, $C_{\nu+\varepsilon}(x)$ normalized so as to be ositive, to

$$\text{(3.3)} \qquad \frac{C'_\nu(x)}{C_\nu(x)} < \frac{C'_{\nu+\varepsilon}(x)}{C_{\nu+\varepsilon}(x)}, \quad c_{\nu+\varepsilon,k-1} < x < c_{\nu k}, \ \varepsilon > 0, \quad \nu \geqq 0,$$

here $c_{\nu k}$ is the k-th positive zero of $C_\nu(x)$.

To establish, say, (3.2), let $g(x) = x$, $f(x) = x^{-1}[x^2 - (\nu+\varepsilon)^2]$, $F(x) = x^{-1}(x^2-\nu^2)$, $y = J_{\nu+\varepsilon}(x)$, $Y = J_\nu(x)$, $a = 0$, $b = j_{\nu 1} (< j_{\nu+\varepsilon,1})$ and recall $_\nu(x) \sim x^\nu$, as $x \to 0+$.

From (3.2), unless $j'_{\nu+\varepsilon,1} \geqq j_{\nu 1}$, it follows that $J'_\nu(j'_{\nu+\varepsilon,1}) < 0$, $\nu \geqq 0$. Thus,

$$\text{(3.4)} \qquad j'_{\nu 1} < j'_{\nu+\varepsilon,1}, \quad \nu \geqq 0, \quad \varepsilon > 0,$$

hich again can be extended to higher ranks. Here as usual, $j'_{\nu k}$ is the k-th positive ero of $J'_\nu(x)$. Below, $c'_{\nu k}$ will denote the k-th positive zero of $C'_\nu(x)$.

From the familiar formula $[x^\nu J_\nu(x)]' = x^\nu J_{\nu-1}(x)$,

$$\frac{J_{\nu-1}(x)}{J_\nu(x)} = \frac{[x^\nu J_\nu(x)]'}{x^\nu J_\nu(x)} = \frac{J'_\nu(x)}{J_\nu(x)} + \frac{\nu}{x},$$

o that its left member can be seen, from (3.2), to be an increasing function of ν, or $\nu > 0$, for each fixed x, $0 < x < j_{\nu 1}$. This provides an alternative proof for hat special case of Lemma 2.3 of M.E.H. Ismail and M.E. Muldoon [3] in which $\beta = 1$, $< x < j_{\nu 1}$, $\nu > -1$. The general case considered in [3] permits $0 < \beta \leqq 1$, and is roved in a different manner.

The corresponding special case namely, that $K_{\nu+1}(x)/K_\nu(x)$ increases with ν, for ach fixed $x > 0$, of their Lemma 2.2 can be established similarly. It suffices to ote that $y = x^{1/2}K_\nu(x)$ is a solution of $y'' + [-\frac{1}{4} + x^{-2}(\frac{1}{4} - \nu^2)]y = 0$, to obtain n analogue of (3.2) and then to continue as above.

In passing, let me thank the authors of [3] for making their manuscript vailable to me and for calling to my attention [7] in which the ratio $J'_\nu(x)/J_\nu(x)$ lays a role, together with $Y_\nu(x)/J_\nu(x)$.

Proofs of (3.1) and (3.4), based on quite different and less elementary reasoning, are found in [12, §15.6]. However, those proofs establish also that $c_{\nu k}$ and $c'_{\nu k}$ increase with ν, for <u>any</u> ν, not only for $\nu \geqq 0$, a double generalization

In turn, these results cast further light on the role of (2.1): <u>Theorem 2.2 becomes false if hypothesis (2.1) be deleted</u>. To see this, let again f, F, g, y and Y be as in the proof of (3.1), but with $-\frac{1}{2} < \nu < 0$, $\varepsilon > 0$, $2\nu+\varepsilon < 0$. Here, $F(x) < f(x)$, $Y(0) = Y(j_{\nu 1}) = 0$. If Theorem 2.2 were true without (2.1), it would imply that $j_{\nu+\varepsilon,1} < j_{\nu 1}$, a contradiction.

Thus, the Szegö end condition cannot be completely eliminated.

It can, however, be moderated. The theorems remain valid if (2.1) is weakened to

$$(3.5) \qquad \limsup_{x \to a+} g(x)\{y'(x)Y(x) - y(x)Y'(x)\} \geqq 0 .$$

This is the case because <u>(3.5) implies (2.1), when g, f, F, y and Y satisfy the remaining hypotheses of Theorem 2.1</u>.

From Green's formula (or directly from the differential equations), these conditions imply that $g\{y'Y - yY'\}$ decreases as x decreases to $a+$, and so (2.1) must hold. Thus, the scope of the theorems is not extended by this mild weakening of (2.1) to (3.5).

It is well known [10, Theorem 3.3.2, p. 46] that, in contrast to (3.1), the first positive zero of one of the classical orthogonal polynomials <u>decreases</u> as the degree increases. For, say, the Jacobi and (generalized) Laguerre polynomials, $\alpha > -1$, this follows from Theorem 2.2, since the weight functions involved increase as n increases. Another result is that the graph of the Laguerre polynomial $L_n^{(\alpha)}(x)$ from the origin to the first zero of $L_{n+1}^{(\alpha)}(x)$ lies above that of the latter function. Presumably, inferences could be drawn about solutions of the pertinent equations for non-integer values of the parameter and for other solutions of the given equations, in respect of the various properties discussed in this note.

4. <u>Simultaneous use of two monotonicities</u>.

For these illustrations, $g(x) = 1$ so that the equation actually considered is

$$(4.1) \qquad y'' + \varphi_\nu(x)y = 0, \quad a < x < b ,$$

with $\varphi_\nu(x)$ continuous in x for each relevant ν, monotonic in $x[\nu]$ for fixed $\nu[x]$. Szegö and Turán took these both into account for (4.1) satisfied by $y = (1-x^2)^{\lambda/2 + 1/4} P_n^{(\lambda)}(x)$, $0 < \lambda < 1$.

Their procedure can be illustrated by establishing similar inequalities for the eros of Laguerre polynomials $L_n^{(\alpha)}(x)$, $-1 < \alpha \leqq 1$. Let

$$\varphi_n(x) = \frac{n+(\alpha+1)/2}{x} + \frac{1-\alpha^2}{4x^2} - \frac{1}{4},$$

$$y = y_n(x) = e^{-x/2}x^{(\alpha+1)/2}L_n^{(\alpha)}(x), \quad a = 0,$$

ith x_{nk} the k-th positive zero (increasing order) of $L_n^{(\alpha)}(x)$. Then, $_{n+1,k} < x_{nk}$, $k = 1,2,\ldots,n$. Thus, one may define $h = x_{nk} - x_{n+1,k} > 0$, so that $_n(x) < \varphi_{n+1}(x-h)$. These weight functions give equations with solutions $y_n(x)$, $_{n+1}(x-h)$. Both vanish at $x_{nk} = x_{n+1,k} + h$. The next zero of $y_{n+1}(x-h)$, $x_{n+1,k+1} + h$, ust, by Theorem 2.2, precede the next zero of $y_n(x)$. Hence

4.2) $$x_{n+1,k+1} - x_{n+1,k} < x_{n,k+1} - x_{nk}, \quad k = 1,2,\ldots,n,$$

aralleling Theorems II and III of [11].

Both in [11] and above, it was essential that the (positive) zeros of fixed ank of $P_n^{(\lambda)}(x)$ and $L_n^{(\alpha)}(x)$, respectively, decrease as n increases. This occurs n all systems of orthogonal polynomials [10, Theorem 3.3.2, p. 46], but for, say, essel functions, each zero increases with the order. This associates with each ombination of monotonicities for $\varphi_\nu(x)$, in ν and in x, perhaps two cases.

In the important example of Bessel functions, one equation is

4.3) $$y'' + (e^{2x} - \nu^2)y = 0, \quad y = C_\nu(e^x).$$

ere $\varphi_{\nu+\varepsilon}(x-h) < \varphi_\nu(x)$, for $\varepsilon \geqq 0$, $h \geqq 0$, $\varepsilon + h > 0$, $\nu \geqq 0$.

Let $C_\nu(x)$ and $Z_\nu(x)$ be Bessel functions, linearly independent or not, with espective positive zeros $\gamma_{\nu k}$, $z_{\nu k}$, where (for a fixed pair of positive integers , m) $z_{\nu+\varepsilon,k} < \gamma_{\nu m}$, for whatever $\varepsilon \geqq 0$ may be under discussion, not necessarily ll nonnegative ε.

For this pair k,m of positive integers and for the ε under consideration, $= \log\gamma_{\nu m} - \log z_{\nu+\varepsilon,k} > 0$, so that $\varphi_{\nu+\varepsilon}(x-h) < \varphi_\nu(x)$, $\nu \geqq 0$, and $C_\nu(x)$, $_{\nu+\varepsilon}(x-h)$ are solutions of (4.3) as it stands, and with ν replaced by $\nu+\varepsilon$, x by -h, respectively. Moreover, $C_\nu(e^x)$ and $C_{\nu+\varepsilon}(e^{x-h})$ both vanish at $x = \log\gamma_{\nu m} = \log z_{\nu+\varepsilon,k} + h$. The next zero of $C_\nu(e^x)$ precedes, by Sturm's theory, the next zero of $C_{\nu+\varepsilon}(e^{x-h})$. Hence $\log\gamma_{\nu,m+1} < \log z_{\nu+\varepsilon,k+1} + h$ so that

4.4) $$\frac{\gamma_{\nu,m+1}}{\gamma_{\nu m}} < \frac{z_{\nu+\varepsilon,k+1}}{z_{\nu+\varepsilon,k}}, \quad \text{for } z_{\nu+\varepsilon,k} < \gamma_{\nu m}, \quad \nu \geqq 0.$$

Some special cases may be worth listing:

(4.5) $$\frac{c_{\nu,k+2}}{c_{\nu,k+1}} < \frac{j_{\nu,k+1}}{j_{\nu k}} < \frac{c_{\nu,k+1}}{c_{\nu k}}, \quad k = 1,2,\ldots,\nu \geqq 0,$$

where the second inequality holds only when $J_\nu(x)$ and $C_\nu(x)$ are linearly independent (becoming an equality otherwise); the first holds always.

To verify (4.5), put $\varepsilon = 0$, and note that $c_{\nu k} < j_{\nu k}$ for $\nu \geqq 0$, when $C_\nu(x)$ and $J_\nu(x)$ are linearly independent [4, §5(i), p. 364], a result implicit in [1, p. 211] as well. The first inequality then follows on putting $m = k+1$, $\gamma_{\nu m} = c_{\nu m}$, $z_{\nu k} = j_{\nu k}$ in (4.4), bearing in mind that its conditions hold even when $c_{\nu k} = j_{\nu k}$. The second inequality follows from the choices $m = k$, $\gamma_{\nu k} = j_{\nu k}$, $z_{\nu k} = c_{\nu k}$.

Also,

(4.6) $$\frac{c_{\nu,k+1}}{c_{\nu k}} < \frac{c_{\nu+\varepsilon,k}}{c_{\nu+\varepsilon,k-1}}, \quad 0 \leqq \varepsilon \leqq 1, \quad \nu \geqq 0, \quad k = 2,3,\ldots;$$

this follows from (4.4) on putting $\gamma_{\nu m} = c_{\nu k}$, $z_{\nu+\varepsilon,k+1} = c_{\nu+\varepsilon,k}$, since [12, p. 480] $c_{\nu+\varepsilon,k-1} < c_{\nu k}$ for $0 \leqq \varepsilon \leqq 1$. For $c_{\nu k} = j_{\nu k}$, this interlacing has been established for $0 \leqq \varepsilon \leqq 2$ [12, p. 480] so that

(4.7) $$\frac{j_{\nu,k+1}}{j_{\nu k}} < \frac{j_{\nu+\varepsilon,k}}{j_{\nu+\varepsilon,k-1}}, \quad 0 \leqq \varepsilon \leqq 2, \quad \nu \geqq 0, \quad k = 2,3,\ldots .$$

It should be noticed that (4.7), and <u>a fortiori</u> (4.6), cannot hold for <u>all</u> ε; the right member approaches 1 as $\varepsilon \to \infty$. Indeed, it can be shown by other methods that $c_{\nu,k+1}/c_{\nu k}$, and hence also $\log c_{\nu,k+1} - \log c_{\nu k}$, <u>decreases</u> as $\nu \to \infty$. This shows that it would be incorrect to rely solely on the monotonicity properties of $\varphi_\nu(x)$, even when it is known that the zeros increase with the parameter.

If, instead of (4.3), use is made of

(4.8) $$y'' + [1 - x^{-2}(\nu^2 - \tfrac{1}{4})]y = 0, \quad x > 0, \quad y = x^{1/2}C_\nu(x),$$

then information may be gleaned about the differences of the zeros rather than their ratios. The same proof as used for (4.4) yields now, for $\varepsilon \geqq 0$,

(4.9) $$\gamma_{\nu,m+1} - \gamma_{\nu m} < z_{\nu+\varepsilon,k+1} - z_{\nu+\varepsilon,k}, \text{ if } z_{\nu+\varepsilon,k} < \gamma_{\nu m}, \quad \nu > \tfrac{1}{2},$$

nd so, keeping in mind interlacing properties of the zeros and the inequality $_{\nu k} < j_{\nu k}$,

$$c_{\nu,k+1} - c_{\nu k} < c_{\nu+\varepsilon,k} - c_{\nu+\varepsilon,k-1}, \quad k = 2,3,\ldots,\ 0 \leqq \varepsilon \leqq 1,\ \ \nu > \frac{1}{2}, \tag{4.10}$$

nd

$$j_{\nu,k+1} - j_{\nu k} < c_{\nu,k+1} - c_{\nu k} < j_{\nu k} - j_{\nu,k-1}, \quad k = 2,3,\ldots,\ \nu > \frac{1}{2}. \tag{4.11}$$

A numerical check of (4.11) gives some idea of the precision of the inequalities, suggesting that they may be of some use in checking tables. Let $\nu = 20.5$, $= 49$, and $C_\nu(x) = Y_\nu(x)$.

Substitution of numerical values gives $3.16088153 < 3.16121708 < 3.16156152$.

For $0 \leqq \nu < \frac{1}{2}$, where $\varphi_{\nu+\varepsilon}(x) < \varphi_\nu(x-h)$,

$$\gamma_{\nu,m+1} - \gamma_{\nu m} < z_{\nu+\varepsilon,k+1} - z_{\nu+\varepsilon,k}, \ \text{if}\ \ \gamma_{\nu m} < z_{\nu+\varepsilon,k}, \quad 0 \leqq \nu \leqq \nu+\varepsilon < \frac{1}{2}, \tag{4.12}$$

so that, in particular, for $k = 1,2,\ldots,$

$$\gamma_{\nu,k+1} - \gamma_{\nu k} < c_{\nu+\varepsilon,k+1} - c_{\nu+\varepsilon,k} \quad \text{if}\ \ c_{\nu+\varepsilon,k} > \gamma_{\nu k}; \quad 0 \leqq \nu \leqq \nu+\varepsilon < \frac{1}{2}, \tag{4.13}$$

$$j_{\nu,k+1} - j_{\nu k} < c_{\nu,k+2} - c_{\nu,k+1} < j_{\nu,k+2} - j_{\nu,k+1}, \quad 0 \leqq \nu < \frac{1}{2}, \tag{4.14}$$

which becomes, when $\nu = 1/3$, $k = 38$,

$$3.1415777 < 3.1415781 < 3.1415784\,.$$

For the range $-\frac{1}{2} < \nu \leqq 0$, where $\varphi_\nu(x) < \varphi_{\nu+\varepsilon}(x-h)$,

$$z_{\nu+\varepsilon,k+1} - z_{\nu+\varepsilon,k} < \gamma_{\nu,m+1} - \gamma_{\nu m} \quad \text{if}\ \ z_{\nu+\varepsilon,k} < \gamma_{\nu m}, \tag{4.15}$$

where $-\frac{1}{2} < \nu \leqq \nu+\varepsilon \leqq 0$.

In particular, for $k = 2,3,\ldots,$

$$j_{\nu k} - j_{\nu,k-1} < c_{\nu,k+1} - c_{\nu k} < j_{\nu,k+1} - j_{\nu k}, \quad -\frac{1}{2} < \nu \leqq 0, \tag{4.16}$$

where the notations $c_{\nu k}$ and $\gamma_{\nu k}$ are understood to be governed by the convention that each is a given function of ν (k fixed) which is not altered even if a change in rank occurs due to the possible appearance or disappearance of zeros as ν varie through a certain set of negative values [12, p. 509].

With the same understanding in the remaining range, where $\varphi_{\nu+\varepsilon}(x-h) < \varphi_\nu(x)$, the following inequalities are found, as special cases of (4.9) which holds for $\nu < -\frac{1}{2}$,

(4.17) $$j_{\nu,k+1} - j_{\nu k} < c_{\nu,k+1} - c_{\nu k} < j_{\nu k} - j_{\nu,k-1}, \quad k = 2,3,\ldots, \ \nu < -\frac{1}{2}.$$

Inequalities (4.9) - (4.17) overlap with those established in [5, Theorem 2] by a different method, and, like them, may be of some help in checking tables of zeros of Bessel functions. Inequalities (4.15) and (4.16), dealing with $-\frac{1}{2} < \nu \leqq 0$ have no counterparts in [5].

5. Acknowledgement.

This work was supported by the National Research Council of Canada. I am grateful also for the generous invitation to present it at the Uppsala Conference.

References

1. M. Bôcher, On certain methods of Sturm and their applications to the roots of Bessel's functions, Bull. Amer. Math. Soc., 3, 1897, 205-213.

2. E. L. Ince, Ordinary differential equations, Dover reprint, 1944.

3. M. E. H. Ismail and M. E. Muldoon, Monotonicity of the zeros of a cross-product of Bessel Functions, SIAM J. Math. Analysis, to appear.

4. L. Lorch and D. J. Newman, A supplement to the Sturm comparison theorem, with applications, Amer. Math. Monthly, 72, 1965, 359-366; Acknowledgement of priority, ibid., 980.

5. L. Lorch and P. Szego, Monotonicity of the differences of zeros of Bessel functions as a function of order, Proc. Amer. Math. Soc., 15, 1964, 91-96.

6. E. Makai, On a monotonic property of certain Sturm-Liouville functions, Acta Math. Acad. Sci. Hungar., 3, 1952, 165-172.

7. L. Z. Salchev and V. B. Popov, Determination of the zeros of a cross-product Bessel function, Proc. Cambridge Philos. Soc., 74, 1973, 477-483.

8. Ch. Sturm, Sur les équations différentielles linéaires du second ordre, J. Math. Pures et Appl., 1, 1836, 106-186.

9. C. A. Swanson, Comparison and oscillation theory of linear differential equations, Academic Press, 1968.

0. G. Szegö, Orthogonal polynomials, 4th ed., Amer. Math. Soc., 1975.

1. G. Szegö and P. Turán, On the monotone convergence of certain Riemann sums, Publ. Math. Debrecen, 8, 1961, 326-335.

2. G. N. Watson, A treatise on the theory of Bessel functions, 2nd ed., Cambridge University Press, 1944.

ork University
ownsview, Ontario, Canada

ostscript: (a) In §1, mention was made of Makai's use of Sturm methods to prove that he areas of arches (from one zero to the next) of the graphs of certain functions form monotonic sequences. Defining "half-arches" as the portions of the graph of a function determined by a zero and an adjacent preceding or succeeding extremum, P. Hartman and A. Wintner [On nonconservative linear oscillators of low frequency, Amer. J. Math., 70, 1948, 529-539] showed that the sequence of areas of half-arches of any solution of (4.1) with $\varphi(x)$ monotonic in x (ν is irrelevant) is also monotonic. I. Bihari [Oscillation and monotonity theorems concerning non-linear differential equations of the second order, Acta Math. Acad. Sci. Hungar., 9, 1958, 83-104] showed the same to be true for a class of equations substantially broader than the Sturm-Liouville equations.

(b) In publications not available to me at the time of this writing, T. A. Sarymsakov discussed the distribution of zeros of solutions of (4.1). References can be found in articles published about him in Uspekhi Mat. Nauk, 21, 1966, 248-253 [= Russian Math. Surveys, 21, 1966, 225-228] and Uspekhi Mat. Nauk, 31, 1976, 241-246 [= Russian Math. Surveys, 31, 1976, 215-221].

W.A. Marcenko

SPEKTRALTHEORIE UND NICHTLINEARE GLEICHUNGEN

Es hat alles im Jahre 1967 begonnen, als C.S. Gardner, I.M. Greene, M.D. Kruskal, R.M. Miura [1] die Lösung des Cauchy-Problems für die KdV Gleichung

$$\dot{v}_t = 6 v v'_x - v'''_{xxx} \quad , \quad v(x,0) = v_0(x)$$

in der Klasse der im Unendlichen verschwindenden Funktionen ($\lim_{x \to \pm\infty} v(x,t) = 0$) mit Hilfe des inversen Problems der Streutheorie für die Sturm-Liuvillischen Operatoren

$$-\frac{d^2}{dx^2} + v(x,t)$$

erhalten haben. Im Jahre 1974 haben P.Lax [2] ,W.A.Marcenko [4], S.P. Novikov [5] gezeigt, wie man auch periodische ($v(x+l,t) = v(x,t)$) Lösungen der KdV Gleichung finden kann. Leider habe ich nicht Zeit genug, um die Hauptideen dieser Arbeiten mitzuteilen. Aber ich meine, daß die mit ihnen verbundenen Fragen, die zur Spektraltheorie gehören, mehr den Hauptthemen unserer Konferenz nahe stehen. Und mir ist es auch sehr angenehm, hier in Schweden diese Probleme zu besprechen, da die ersten wichtigen Resultate auf diesem Gebiet von Prof. 'Ambarzumjan und Prof. Borg erreicht wurden.

Betrachten wir den Hillschen Operator

$$H = -\frac{d^2}{dx^2} + q(x) \qquad (-\infty < x < \infty)$$

$$\Im m\, q(x) = 0 \quad , \quad q(x+\pi) = q(x) ,$$

wo $q(x)$ eine reelle und periodische Funktion ist. Die Funktion $q(x)$ werden wir Potential nennen, da H der eindimensionale Schrödingersche Operator mit Potential $q(x)$ ist. Es ist bekannt, daß das Spektrum dieses Operators stetig ist und aus Segmenten $[\mu_0, \mu_1^-], [\mu_1^+, \mu_2^-], [\mu_2^+, \mu_3^-], \dots$

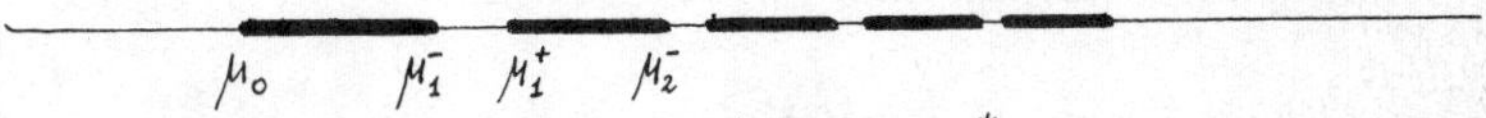

besteht. Hier sind $\mu_0 < \mu_2^- \leq \mu_2^+ < \mu_4^- \leq \mu_4^+ < \dots$
die Eigenwerte der periodischen Randwertaufgabe:

$$H[y] = \mu y \quad (0 \leq x \leq \pi), \quad y(0) - y(\pi) = y'(0) - y'(\pi) = 0 ,$$

und $\mu_1^- \leq \mu_1^+ < \mu_3^- \leq \mu_3^+ < \dots$ die Eigenwerte der antiperiodischen Randwertaufgabe

$$H[y] = \mu y \quad (0 \leq x \leq \pi), \quad y(0) + y(\pi) = y'(0) + y'(\pi) = 0$$

Es war wichtig, die notwendigen und hinreichenden Bedingungen dafür zu finden, daß es für eine Folge $-\infty < a_0 < a_1^- \leq a_1^+ < a_2^- \leq a_2^+ < \dots$ einen solchen Operator H gibt, daß $a_0 = \mu_0$, $a_k^\pm = \mu_k^\pm$ $(k = 1, 2, \dots)$ ist. Das ist das erste Problem, das ich besprechen möchte.

Es gibt ja unendlich viele Potentiale $q(x)$ mit denselben Eigenwerten $\mu_k^\pm$, denn, zum Beispiel, alle Hillsche Operatoren $H_\alpha = -\frac{d^2}{dx^2} + q(x+\alpha)$ haben dasselbe Spektrum. Deshalb muß man bei der Lösung dieses Problems nicht nur die notwendigen und hinreichenden Bedingungen finden, sondern auch alle solche Potentiale beschreiben und eine Methode zeigen, durch welche man sie alle erhält.

Dieses Problem hat einige spezifische Schwierigkeiten im V
gleich zu anderen Umkehrproblemen der Spektraltheorie.

Zum Beispiel, wenn die Folgen $\{\lambda_k\}, \{\nu_k\}$ aus Eigenwerten der Randwertprobleme

$$H[y] = \lambda y \quad (0 \le x \le \pi), \quad y(0) = y(\pi) = 0,$$

$$H[y] = \nu y \quad (0 \le x \le \pi), \quad y(0) = y'(\pi) = 0$$

bestehen, dann gibt es solche offene Intervale $(\lambda_k - \varepsilon_k, \lambda_k + \varepsilon$ $(\nu_k - \varepsilon_k, \nu_k + \varepsilon_k)$ $(\varepsilon_k > 0)$, daß jede beliebige Folge $\tilde{\lambda}_k \in$ $\in (\lambda_k - \varepsilon_k, \lambda_k + \varepsilon_k)$, $\tilde{\nu}_k \in (\nu_k - \varepsilon_k, \nu_k + \varepsilon_k)$ auch aus Eigenwerten von analogen Randwertproblemen

$$\tilde{H}[y] = \tilde{\lambda} y \quad (0 \le x \le \pi), \quad y(0) = y(\pi) = 0,$$

$$\tilde{H}[y] = \tilde{\nu} y \quad (0 \le x \le \pi), \quad y(0) = y'(\pi) = 0$$

besteht. Also, die Folgen der Eigenwerte dieser Randwertprobleme bilden eine offene Menge in einem gewissen Folgenraum.
Anders ist es mit Folgen von Eigenwerten der periodischen und
antiperiodischen Randwertaufgaben. Wenn sich ein Eigenwert
ändert, so ändern sich auch alle anderen. Damit bilden die
möglichen Eigenwertfolgen eine gewisse Mannigfaltigkeit in de
genannten Folgenraum. Das Problem besteht also in der Paramet
risierung dieser Mannigfaltigkeit durch unabhängige Parameter

Ich muß noch auf die folgende Tatsache aufmerksam machen:
es gibt solche Potentiale $q(x)$, daß die Ungleichung

$$\mu_k^- \neq \mu_k^+$$

nur für eine endliche Anzahl von Indices $k = k_1, k_2, \dots, k_{N-1}$ gilt. Das heißt, $\mu_k^- = \mu_k^+$ wenn k von $k_1, k_2, \dots, k_{N-1}$ verschieden ist. Solche Potentiale nennen wir Potentialen mit N Zonen, denn das Spektrum des Hillschen Operators besteht in diesem Fall nur aus N Zonen:

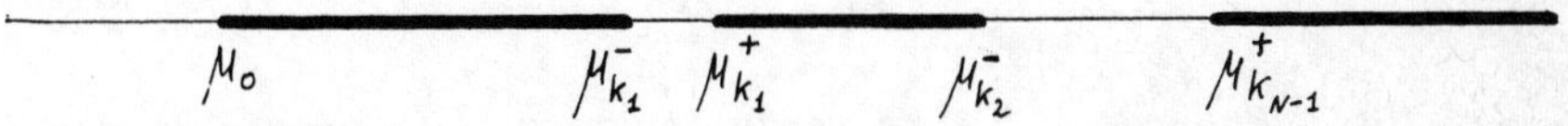

Bezeichnen wir die Menge solcher Potentiale durch K_N und es sei K - die Vereinigung der K_N : $K = \bigcup_N K_N$. So kommen wir zu folgenden Problemen:

2) Ist es möglich, ein beliebiges reelles periodisches Potential $q(x)$ mit Potentialen von K zu approximieren:

$$q(x) = \lim_{N\to\infty} q_N(x) \quad (q_N(x) \in K_N) \quad ?$$

3) Wie hängt die Geschwindigkeit, mit der die Folge der Abstände von $q(x)$ bis K_N

$$\rho(q, K_N) = \inf_{q_N \in K_N} \| q(x) - q_N(x) \|$$

nach Null hinstrebt von den Eigenschaften vom Potential $q(x)$ ab? (Hier kann die Norm $\| q(x) - q_N(x) \|$ auf verschiedene Art definiert werden).

Um die Resultate, die wir mit Prof. I.W.Ostrovskij [6] erreicht haben, formulieren zu können, führen wir die folgenden Bezeichnungen ein:

$C(\lambda, x)$, $S(\lambda, x)$ sind die Lösungen der Gleichung $H[y] = \lambda^2 y$ die durch die Anfangsbedingungen $C(\lambda, 0) = S'(\lambda, 0) = 1$, $C'(\lambda, 0) = S(\lambda, 0) = 0$ bestimmt werden, $U_+(\lambda) = \frac{1}{2}\left[C(\lambda, \pi) + S'(\lambda, \pi)\right]$ ist die Ljapunovsche Funktion oder Hillsche Diskriminante. Außerdem sei

$$U_-(\lambda) = \frac{1}{2}\left[C(\lambda, \pi) - S'(\lambda, \pi)\right].$$

Es ist bekannt, daß $\mu_k^{\pm}$ die Wurzeln der Gleichung $u_+^2(\sqrt{z})-1=0$ sind. Und es ergibt sich daraus, daß die ga Funktion $u_+(\sqrt{z^2+\mu_0})$ folgende Eigenschaften hat: sie ist ree ($\operatorname{Im} u_+(\sqrt{x^2+\mu_0})=0$, $-\infty<x<\infty$) und alle Wurzeln de Gleichung $u_+^2(\sqrt{z^2+\mu_0})-1=0$ sind reell ($z_0=0$, $z_k^{\pm}=\sqrt{\mu_k^{\pm}-\mu_0}$, $z_{-k}^{\pm}=-\sqrt{\mu_k^{\mp}-\mu_0}$, $k=1,2,\ldots$).

Theorem 1. Eine ganze Funktion $u(z)$ erfüllt die Bedingungen

A) $\operatorname{Im} u(x)=0$, $-\infty<x<\infty$,

B) $u^2(z_\nu)-1=0 \Rightarrow \operatorname{Im} z_\nu=0$

dann, und nur dann, wenn sie durch die Formel

$$u(z)=\cos\Theta(z)$$

dargestellt wird. Hier ist

$$\Theta(z)=\begin{cases}\Theta_+(z) & \operatorname{Im} z\geq 0\\ \overline{\Theta_+(\bar z)} & \operatorname{Im} z<0\end{cases},$$

wo $\Theta_+(z)$ eine komplexe Funktion ist, die die Halbebene $Z^+=\{z:\operatorname{Im} z>0\}$ auf das Gebiet

$$\Theta^+\{h_k\}=\{\theta: p\pi<\operatorname{Re}\theta<q\pi,\ \operatorname{Im}\theta>0\}\setminus$$
$$\setminus\bigcup_{p<k<q}\{\theta:\operatorname{Re}\theta=k\pi,\ 0<\operatorname{Im}\theta\leq h_k\}$$

konform abbildet.

Hier ist der Fall abgebildet, wenn $-\infty < p < q = +\infty$ **ist:**

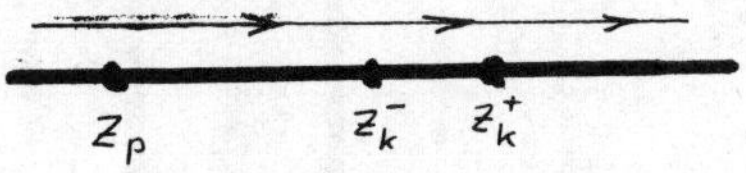

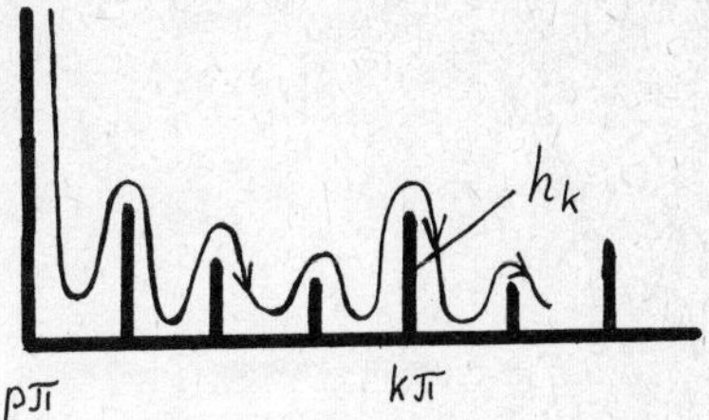

$u^2(z_k^\pm) - 1 = 0$ $\qquad -\infty < p < q = +\infty .$

Die Funktion $\theta(z)$ **ist eine analytische Fortsetzung der Funktion** $\theta_+(z)$ **, welche das Gebiet**

$$Z\{z_k^\pm\} = \{z\} \setminus \Big(\{z : \operatorname{Im} z = 0,\ \operatorname{Re} z \le z_p\} \cup \{z : \operatorname{Im} z = 0,\ \operatorname{Re} z \ge z_q\} \bigcup_{p<k<q} \{z : \operatorname{Im} z = 0,\ z_k^- \le \operatorname{Re} z \le z_k^+\} \Big)$$

auf das Gebiet

$$\Theta\{h_k\} = \{\theta : p\pi < \operatorname{Re}\theta < q\pi\} \setminus \bigcup_{p<k<q} \{\theta : \operatorname{Re}\theta = k\pi,\ -h_k \le \operatorname{Im}\theta \le h_k\}$$

konform abbildet:

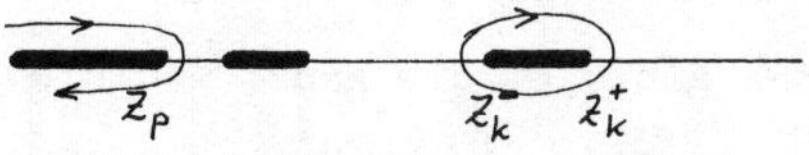

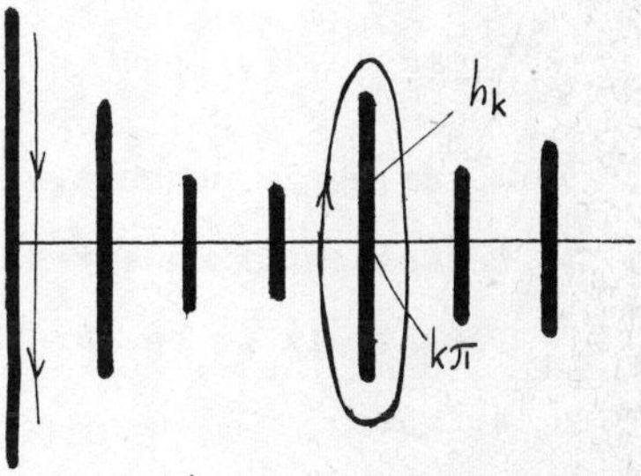

Da die Ljapunovsche Funktion $u_+(\sqrt{z^2+\mu_0})$ **die Bedingungen A), B) erfüllt, kann sie auch durch die Formel** $u_+(\sqrt{z^2+\mu_0}) = \cos\theta(z)$ **dargestellt sein. Und in diesem Fall ist** $p = -\infty$, $q = +\infty$,

$$h_0 = 0\ ,\quad h_k = h_{-k}\ ,\quad \theta(0) = 0\ ,\quad \lim_{y \to +\infty} (iy)^{-1}\theta(iy) = \pi\ :$$

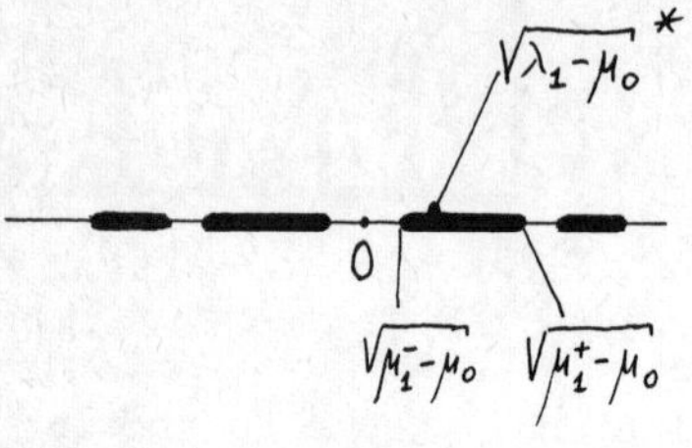

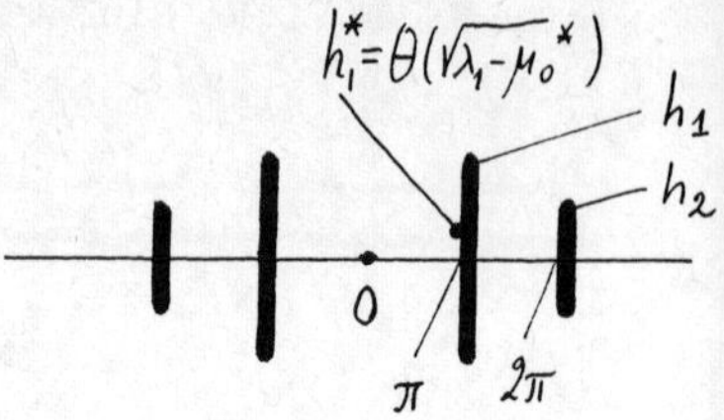

Die Größe dieser Schlitze $2h_k$ hängt von der Differentierbarkeit des Potentials $q(x)$ ab. Es sei

$$\widetilde{W}_2^n = \left\{ q(x) : q(x) = \sum_{m=-\infty}^{\infty} q_m e^{imx}, \quad q_m = \overline{q_{-m}}, \quad \|q\|_2^n = \left[q_0^2 + \sum_{m=-\infty}^{\infty} |m^n q_m|^2 \right]^{\frac{1}{2}} < \infty \right\}.$$

Dann gilt folgendes

<u>Lemma</u> $q(x) \in \widetilde{W}_2^n \Rightarrow \sum_{k=1}^{\infty} (k^{n+1} h_k)^2 < \infty .$

Es wurde schon erwähnt, daß es viele Potentiale $q(x)$ mit denselben Eigenwerten $\mu_k^{\pm}$ gibt. Um die ganze Menge solcher Potentiale zu beschreiben, betrachten wir noch eine Randwertaufgabe

$$H[y] = \lambda y \quad (0 \le x \le \pi), \quad y(0) = y(\pi) = 0 \tag{1}$$

Wir bezeichnen mit λ_k ihre Eigenwerte. Es ist bekannt, daß immer die Ungleichungen $\mu_k^- \le \lambda_k \le \mu_k^+$ erfüllt sind. Aus einem Eindeutigkeitstheorem von Prof. G. Borg [3] folgt, daß die Folgen

$$\left.\begin{array}{l} \mu_0 < \mu_1^- \le \lambda_1 \le \mu_1^+ < \mu_2^- \le \lambda_2 \le \mu_2^+ < \dots \\ \operatorname{sign} u_-(\sqrt{\lambda_1}), \operatorname{sign} u_-(\sqrt{\lambda_2}), \dots \end{array}\right\} \quad (2)$$

das Potential $q(x)$ eindeutig bestimmen.

Definieren wir die Folge $h_k \ge 0$ durch

$$\operatorname{ch} h_k = \max_{\mu_k^- \le x \le \mu_k^+} |u_+(\sqrt{x})|$$

und betrachten wir das Gebiet $\Theta\{h_k\}$ $(h_{-k} = h_k)$.

Es liege $\sqrt{\lambda_k - \mu_0} \in [\sqrt{\mu_k^- - \mu_0}, \sqrt{\mu_k^+ - \mu_0}]$ auf dem oberen (unteren) Ufer des Schlitzes $\{z : \operatorname{Im} z = 0, \sqrt{\mu_k^- - \mu_0} \le \operatorname{Re} z \le \le \sqrt{\mu_k^+ - \mu_0}\}$ wenn $u_-(\sqrt{\lambda_k}) \ge 0$ $(u_-(\sqrt{\lambda_k}) < 0)$ ist, und bezeichnen wir den entsprechenden Punkt mit $\sqrt{\lambda_k - \mu_0}^*$. Auf diese Weise bekommen wir eine eineindeutige Beziehung zwischen Folgen (2) und den zugehörigen

$$\mu_0; h_1, h_2, \dots \quad ; h_1^* = \theta(\sqrt{\lambda_1 - \mu_0}^*), h_2^* = \theta(\sqrt{\lambda_2 - \mu_0}^*), \dots$$

und dabei ist

$$\mu_k^\pm = \mu_0 + z^2(k\pi \pm 0), \quad \lambda_k = \mu_0 + z^2(h_k^*),$$
$$\operatorname{sign} u_-(\sqrt{\lambda_k}) = \operatorname{sign} \operatorname{Im} h_k^*.$$

Hier ist mit $Z(\theta)$ eine inverse Funktion von $\theta(Z)$ bezeichnet.

Aus dem Lemma folgt, daß $\sum_{k=1}^{\infty}\left(k^{n+1}h_k\right)^2<\infty$, wenn $q(x)\in\widetilde{W}_2^n$.

Theorem 2. Es sei $\{h_k\}$ eine beliebige Folge mit $h_0=0$, $h_k=h$ $\sum_{k=1}^{\infty}(k^{n+1}h_k)^2<\infty$, μ_0 - eine beliebige reelle Zahl, - eine beliebige Folge von Punkten auf den Ufern der Schlitze $\{\theta : Re\,\theta=k\pi, -h_k\leq Im\,\theta\leq h_k\}$. Dann gibt es ein und nur ein Potential $q(x)\in\widetilde{W}_2^n$, so daß für den Operator $H=-\frac{d^2}{dx^2}+q(x)$ die Eigenwerte der periodischen und antiperiodischen Randwertaufgaben $\mu_k^{\pm}=\mu_0+Z^2(k\pi\pm 0)$ sind, die Eigenwerte der Randwertaufgabe (1) $\lambda_k=\mu_0+Z^2(h_k^*)$ sind und die Gleichungen $sign\,U_-(\sqrt{\lambda_k})=sign\,Im\,h_k^*$ gelten.

Dieses ist die Lösung des ersten Problems und die Formel

$$\mu_k^{\pm}=\mu_0+Z^2(k\pi\pm 0)$$

geben die Parametrisierung der obengenannten Mannigfaltigkeit durch unabhängige Parameter $\mu_0; h_1, h_2, \dots$.

Das zweite und auch das dritte Problem läßt sich folgendermaßen lösen.

Es sei ein Potential $q(x)\in\widetilde{W}_2^n$ gegeben. Ihm entspricht eine Folge $\mu_0; h_1, h_2, \dots; h_1^*, h_2^*, \dots$. Schneiden wir die Folge h_k nach dem N-ten Gliede ab, so erhält man die Folge:

$$\mu_0; h_1, h_2, \dots h_N, 0, 0, \dots; h_1^*, h_2^*, \dots h_N^*, (N+1)\pi, (N+2)\pi, \dots$$

Dazu gehört wieder ein Potential $q_{N+1}(x)$ und dieses hat höchstens $N+1$ Zonen.

Nun dann gilt folgendes:

Theorem 3.

$$\|q(x)-q_{N+1}(x)\|_2^n \leq C\left\{\sum_{k>N}(k^{n+1}h_k)^2\right\}^{\frac{1}{2}} \exp C\left\{\sum_{k=1}^{\infty}(k^{n+1}h_k)^2\right\}^{\frac{1}{2}} \leq$$

$$\leq C_1\left\{\sum_{k=N+1}^{\infty}(k^n\Delta_k)^2\right\}^{\frac{1}{2}} \exp C_1\|q\|_2^n ,$$

wobei $C=C(n)$ und $C_1=C_1(n)$ nur von n abhängen und $\Delta_k=\mu_k^+-\mu_k^-$ ist.

Also

$$\rho(q,K_N)=\inf_{q_N\in K_N}\|q(x)-q_N(x)\|_2^n \leq C_2\left\{\sum_{k=N}^{\infty}(k^n\Delta_k)^2\right\}^{\frac{1}{2}}$$

wobei $C_2=C_2(n,q)$ nur von n und $\|q\|_2^n$ abhängt.

So sind alle drei Probleme gelöst.

Die Verallgemeinerung dieser Probleme auf Systeme ist bisher nur für den Fall des Diracschen Operators

$$\mathcal{D}=\begin{pmatrix}0 & 1\\ -1 & 0\end{pmatrix}\frac{d}{dx}+\begin{pmatrix}p(x) & r(x)\\ r(x) & -p(x)\end{pmatrix}$$

durchgeführt worden (T. Mißüra).

Es wäre sehr interessant, auch andere Fälle zu behandeln.

1.C.S. Gardner, I.M. Greene, M.D. Kruskal, R.M. Miura, Phys. Rev. Lett., 19 (1967), 1095-1097. 2. P.D. Lax, Lect. in Appl. Math., 15 (1974), 85-96. 3. G. Borg, Acta Math., 78:1 (1946), 1-96. 4. В.А.Марченко, Матем. сб., 95 / I37 /, / I974 /,33I-356.

5. С.П.Новиков, Функц. анализ, 8:3 /I974/, 54 - 66.

6. В.АМарченко, И.В.Островский, Матем. сб., 97 /I39/,/I975/, 540 - 606.

On a class of polynomials connected with the Korteweg - de Vries equation

J. Moser *)

1. Introduction.

This note reports on some recent results by M. Adler and the author which wi appear in full in a Math. Research report of the University of Wisconsin. This wor extends [1].

There is an intimate relation between the spectral theory of the Sturm-Liouville operators

(1) $$L = -D^2 + u \; ; \quad D = \frac{d}{dx}$$

and the partial differential equation

(2) $$u_t = 3uu_x - \frac{1}{2}u_{xxx},$$

the Korteweg - de Vries equation. This connection is provided by the remarkable fac due to Kruskal and P.D. Lax, that the solutions $u = u(x,t)$ of (2) give rise tc operators $L = L_t$ whose spectrum is independent of t, i.e. they generate "iso-spectral deformations" of $L_0 = -D^2 + u(x,0)$. To make this statement meaningful one has to impose appropriate boundary conditions and restrict the class of potentials suitably. For example, for L defined on the whole real axis one requires

$$\int_{-\infty}^{+\infty} (1 + |x|)|u|dx < \infty$$

so that the symmetric operator L on C''_{comp} (C''- functions with compact support) has a unique selfadjoint extension (see, e.g. [4]). Another situation which has been studied by many authors is the case of the Hill's equation: We assume $u(x,t) = u(x+1, t)$ and impose periodic boundary conditions $y(0) = y(1)$, $y'(0) = y'(1)$ (see [3, 5, 6]) for the functions y in the domain of L.

---

*) This work was partially supported by the National Science Foundation Grant MCS 74-03003-A-02.

Here we restrict ourselves to a different class of functions, namely the solutions of (2) which are rational in x for all values of t. It turns out that these solutions are automatically rational also in t and it is possible to give rather explicit description in terms of recursion formulae for these potentials. These potentials are closely related to the wellknown operators

$$-D^2 + \frac{d(d+1)}{x^2} \tag{3}$$

whose eigenfunctions are expressed in terms of Bessel-function. However, we will not describe the spectral properties of these rational potentials but mention only that a very special case was investigated by Moses and Tuan [7].

2. The Higher Korteweg - de Vries equations.

To describe our result we have to mention that the partial differential equation (2) is not the only one providing iso-spectral deformations, but there is a whole sequence of such nonlinear differential operators

$$u_t = X_k u \qquad (k = 1,2,\dots) \tag{4}$$

where

$$X_1 u = u_x, \quad X_2 u = 3uu_x - \frac{1}{2} u_{xxx},$$

while X_k is a differential operator of order $2k-1$. According to a formula of A.Lenard (see [4]) the X_k can be recursively defined as follows: Set

$$X_k = \frac{\partial}{\partial x} \frac{\partial H_k}{\partial u} \tag{5}$$

where $H_k = \int P_k(u, u', \dots)dx$ with P_k being polynomials in u and finitely many of its derivatives and

$$\frac{\partial H_k}{\partial u} = \frac{\partial P_k}{\partial u} - \frac{\partial}{\partial x} \frac{\partial P_k}{\partial u'} + \left(\frac{\partial}{\partial x}\right)^2 \frac{\partial P_k}{\partial u''} - + \dots$$

its Lagrange derivative. The recursion formula is given by (5) together with

$$X_{k+1} u = \left(uD + Du - \frac{1}{2} D^3\right) \frac{\partial H_k}{\partial u}. \tag{6}$$

It is known that this recursion is consistent, and defines the X_k up to a linear combination $c_1X_1 + c_2X_2 + \dots c_{k-1}X_{k-1}$. We can normalize the X_k by requiring that $X_k u$ is iso-basic, i.e. of weight $2k+1$ if we assign u the weight 2 an $\frac{\partial}{\partial x}$ the weight 1.

The partial differential equations (4) are sometimes referred to as the higher Korteweg - de Vries equations although they were found only in recent times. The form (5) of these operators suggest that they are Hamiltonian systems of differential equations, H_k being the Hamiltonians. Finally, we point out that these operators commute:

$$X_k X_j = X_j X_k ,$$

and therefore the corresponding flows $\exp(t_k X_k)$ commute also.

3. The polynomials θ_d.

We address ourselves to the question to determine the manifold $\mathcal{M}$ of rational functions of x vanishing at $x = \infty$ which is invariant under all the flows $\exp(t_k X_k)$. According to [1] $\mathcal{M}$ decomposes into denumerably manifolds $\mathcal{M}_d$ of dimension d for $d = 1, 2, \dots$ and, moreover, each $\mathcal{M}_d$ is generated the single function

$$\frac{d(d+1)}{x^2} . \tag{7}$$

One of the main results of this note is the representation of all $u_d \in \mathcal{M}_d$ in the form

$$u_d = -2D^2 \log \theta_d \tag{8}$$

where $\theta_d = \theta_d(x+\tau_1, \tau_2, \dots, \tau_d)$ are polynomials of the d arguments with rational coefficients. These polynomials can be determined by the recursion formula ("prime" stands for differentiation with respect to x)

$$\theta'_{k+1}\,\theta_{k-1} - \theta_{k+1}\,\theta'_{k-1} = (2k+1)\,\theta_k^2 \tag{9}$$

for $k = 1, 2, \dots$ and

$$\theta_0 = 1\ , \quad \theta_1 = x + \tau_1\ .$$

It is remarkable that the equation (9) has as a solution θ_{k+1} a polynomial in x. Of course, the solution depends on an integration constant which can be fixed by requiring that the coefficient of

$$x^{\frac{1}{2}k(k-1)}$$

is equal to τ_{k+1}, thus introducing the parameters $\tau_1, \tau_2, \ldots$.

We list the first few polynomials for $\tau_1 = 0$:

$$\theta_2 = x^3 + \tau_2$$

$$\theta_3 = x^6 + 5\tau_2 x^3 + \tau_3 x - 5\tau_2^2$$

$$\theta_3 = x^{10} + 15\tau_2 x^7 + 7\tau_3 x^5 - 35\tau_2\tau_3 x^2 + 175\tau_2^3 x - \frac{7}{3}\tau_3^2 + \tau_4 x^3 + \tau_4\tau_2\ .$$

These polynomials have a number of additional properties:

i) If $[A,B] = A'B - AB'$ denotes the Wronskian determinant one has

$$[\theta_k', \theta_{k-1}] + [\theta_{k-1}', \theta_k] = 0\ .$$

ii) The polynomials $\theta_k(x, \tau_2, \ldots, \tau_k)$ are of degree $n_k = \frac{1}{2}k(k+1)$ in x and are monic,

$$\theta_k = x^{n_k} + \ldots\ .$$

iii) The homogeneity property

$$\theta_k(\lambda\tau_1, \lambda^3\tau_2, \ldots, \lambda^{2k-1}\tau_k) = \lambda^{n_k}\,\theta_k(\tau_1, \ldots, \tau_k)$$

holds for any complex λ.

iv) For $\tau_1 = \tau_2 = \ldots = \tau_k = 0$ one has

$$\theta_k = x^{n_k}$$

so that the corresponding u_k defined by (8) is

$$u_k = \frac{2n_k}{x^2} = \frac{k(k+1)}{x^2} .$$

4. The main result.

The main property of these polynomials is contained in their relation to the Korteweg - de Vries equation. To explain this relation we observe that there is considerable arbitrariness in the way the parameters $\tau_2, \tau_3, \ldots$ were introduced, and it would have been just as well, instead, to introduce the parameters

$$(10) \qquad \tau_k^* = \alpha_k \tau_k + p_k(\tau_2, \tau_3, \ldots, \tau_{k-1})$$

where the p_k are polynomials and $\alpha_k \neq 0$ rational numbers. To respect the homogeneity property iii) we require

$$p_k(\lambda^3 \tau_2, \lambda^5 \tau_3, \ldots, \lambda^{2k-3} \tau_{k-1}) = \lambda^{2k-1} p_k(\tau_2, \ldots, \tau_{k-1}) .$$

In particular, we find $p_k = 0$ for $k = 2,3,4$, i.e. $\tau_k^* = \alpha_k \tau_k$ for $k = 2,3,4$. We denote the new polynomials by $\theta_k^* = \theta_k^*(x + \tau_1, \tau_2^*, \ldots, \tau_k^*)$.

Statement: With an appropriate choice of the parameters τ_k^* via a birational transformation (10) the functions

$$u_d^* = - 2D^2 \log \theta_d^*$$

are solutions of the Korteweg - de Vries equation

$$\frac{\partial u_d^*}{\partial t} = X_k u_d^*$$

where $t = \tau_k^*$. In particular, since u_d^* is independent of τ_k^* for $k > d$ we have

$$X_k u_d^* = 0 \quad \text{for} \quad k > d .$$

For $k = 1,2,3,4$ we have $\tau_k^* = \alpha_k \tau_k$ and hence

$$\frac{\partial u_d}{\partial \tau_k} = \alpha_k X_k \quad \text{for} \quad k \leq 4 .$$

One computes $\alpha_1 = 1$, $\alpha_2 = -\frac{1}{6}$.

The above statement amounts to the following: Since for $\tau_1 = \tau_2 = \ldots = \tau_d$ the function u_d becomes $d(d+1)x^{-2}$ the functions $u_d(x+\tau_1 \ldots \tau_d)$ are generated by the flow $\exp(\sum_k t_k X_k)$ and this is the manifold $\mathcal{M}_d$ mentioned in Section 3. This explains the remark related to (3) made in the introduction.

5. The eigenfunctions of L.

It is wellknown that the solutions φ of

$$(-D^2 + \frac{d(d+1)}{x^2})\varphi = \lambda\varphi$$

are elementary functions of x if d is an integer. More precisely, if $\lambda = -\omega^2$ they are $e^{\pm\omega x}$ times a rational function of x. The same property holds true for the solutions of

$$(-D^2 + u_d)\varphi = -\omega^2\varphi .$$

To explain this statement as well as to motivate the recursion formula (9) we utilize a wellknown transformation of Sturm-Liouville operators L, which is due to M.M. Crum [2]. We need only a special form: Any second order differential operator $L = -D^2 + u$ can formally be factorized into two first order operator

$$L = A^*A ,$$

where $A = D - v$, $A^* = -(D+v)$ and v is a solution of the Ricatti equation

$$u = v' + v^2 .$$

Alternately, if $\varphi \neq 0$ is a solution of $A\varphi = 0$ then A can be written as

$$A = \varphi D \varphi^{-1} = D - v , \quad v = \frac{\varphi'}{\varphi} ,$$

while

$$A^* = -\varphi^{-1} D \varphi .$$

Now we exchange the rôle of A and A^* and define

$$\tilde{L} = AA^* = -D^2 + \tilde{u}$$

with

$$\tilde{u} = -v' + v^2 .$$

Note that $\tilde{L}$ and L are connected by the intertwining relation

$$\tilde{L}A = AL$$

which shows that

(11) $\quad (L-\lambda)\psi = 0$ implies $(\tilde{L}-\lambda)A\psi = 0$.

This mapping of L into $\tilde{L}$ is a special case of Crum's transformation.

If we apply this transformation to $\tilde{L}$ with φ^{-1} as eigenfunction of $\tilde{L}$ we are led back to L. But the above factorization is not unique and depends on the choice of the eigenfunction φ of $L\varphi = 0$. Thus we are led to a one-parameter family of potentials $\tilde{u}$ by this transformation and by k repetitions to a k-parameter family of potentials. In particular, if we start with $u_0 = 0$ we obtain after applying this transformation k times

$$u_k = -2D^2 \log \theta_k \ ; \quad \varphi_k = \frac{\theta_{k+1}}{\theta_k}$$

$$v_k = \frac{\varphi_k'}{\varphi_k}$$

$$(-D^2 + u_k)\varphi_k = 0 , \quad (-D^2 + u_k)\,\varphi_{k-1}^{-1} = 0 ,$$

where θ_k are the polynomials of Section 3. Therefore $\varphi_k, \varphi_{k-1}^{-1}$ have a constant Wronskian. If we normalize it to be

$$[\varphi_k , \varphi_{k-1}^{-1}] = 2k+1$$

we are led to (9).

Finally, the relation (11) implies that

$$\psi = A_{k-1} \cdots A_1 A_0 \, e^{\pm\omega x}$$

are solutions of $(-D^2 + u_k)\psi = -\omega^2\psi$ if

$$A_j = (D - v_j) = \varphi_j D \varphi_j^{-1} .$$

his shows that ψ equals a rational function of x times $e^{\pm\omega x}$ verifying our laim.

. Explicit Representation of the θ_d.

If one makes use of the formulae of Crum [2] one can represent the polynomials $_d$ as well as the transformation $A_{d-1} \dots A_1 A_0$ in terms of certain Wronskians in xplicit form.

We introduce the Wronskian of k functions $\psi_1, \psi_2, \dots \psi_k$ by

$$W_k = W(\psi_1, \psi_2, \dots \psi_k) = \det(D^{i-1}\psi_j), \quad i,j = 1,2,\dots,k.$$

or abbreviation we also set

$$W_k(\chi) = W(\psi_1,\dots, \psi_k, \chi), \; W_k(\chi,\varphi) = W(\psi_1,\dots, \psi_k, \chi, \varphi)$$

ith two arbitrary smooth functions φ, χ. Then one has Jacobi's identity

(12) $$[W_k(\chi), W_k(\varphi)] = -W_k(\chi,\varphi)\, W_k \quad \text{for } k = 1,2,\dots .$$

e apply the above definitions to a system of functions ψ_j satisfying $\psi_0 = 0$, $\psi_1 = x$ and

(13) $$\psi_j'' = \psi_{j-1} \quad (j = 1,2,\dots, k).$$

Then one verifies that for $\chi = 1$, the Wronskian $W_k(\chi)$ becomes

$$W_k(1) = (-1)^k W_{k-1} \quad (k = 1,2,\dots)$$

if we also set $W_0 = 1$.

Setting $\chi = 1, \varphi = \psi_{k+1}$ in (12) we get

$$[W_{k-1}, W_{k+1}] = [W_k(1), W_k(\psi_{k+1})] = -W_k^2$$

and $W_0 = 1, W_1 = x$. Thus, the W_k satisfy, aside from a constant, the same recursion formula as the θ_k and we conclude that

(14) $$\theta_k = \mu_k W_k$$

with some constants μ_k which one determines to be

$$\mu_k = 1^k \cdot 3^{k-1} \cdot 5^{k-2} \dots (2k-1)! = \prod_{j=1}^{k} (2k-2j+1)^j .$$

The equation (13) with $\psi_1 = x$ gives

$$\psi_j = \frac{x^{2j-1}}{(2j-1)!} + \sum_{i=0}^{j-2} \rho_{j-i} \frac{x^{2i}}{(2i)!}$$

with some integration constants $\rho_2, \rho_3, \dots$. Equivalently we can define the ψ_j via the generating function

$$\sum_{j=0}^{\infty} \psi_j s^{2j-1} = \sinh(sx) + \Big(\sum_{j=2}^{\infty} \rho_j s^{2j-1} \Big) \cosh(sx) .$$

With this choice of $\psi_1, \psi_2 \dots$ the formula (14) gives the desired representation of $\theta_k = \theta_k(x, \tau_2, \dots \tau_k)$ where the $\tau_2, \dots \tau_k$ are birationally and isobarically related to $\rho_2, \rho_3, \dots, \rho_k$. For example one finds $\tau_2 = -3\rho_2$, $\tau_3 = 45\rho_3$ an $\mu_1 = 1$, $\mu_2 = 3$, $\mu_3 = 45$.

The mapping

$$\chi \rightarrow T_k \chi = (A_{k-1}, \dots A_1 A_0)\chi$$

of Section 4 is given by the formula

$$T_k \chi = \frac{W_k(\chi)}{W_k} .$$

Indeed, for $k = 1$ this is clearly the case, and the general case follows from the relation

$$A_k \frac{W_k(\chi)}{W_k} = \varphi_k D \left(\varphi_k^{-1} \frac{W_k(\chi)}{W_k} \right) = \frac{W_{k+1}(\chi)}{W_{k+1}}$$

which follows from $\varphi_k = \dfrac{\theta_{k+1}}{\theta_k}$ and (12).

By our construction we have

$$L_k T_k = T_k(-D^2)$$

and therefore

$$\varphi_{\pm} = \frac{W_k(e^{\pm\omega x})}{W_k}$$

re solutions of $(L_k + \omega^2)\varphi_{\pm} = 0$ whose Wronskian turns out to be

$$[\varphi_+ , \varphi_-] = 2\omega^{2k+1} .$$

The above Wronskian enters in the denominator of the Green's function of $_k + \omega^2$ and we see that the resolvent has a higher order singularity at $\lambda = 0$. We onclude by describing the eigenfunction (generalized) of L_k for $\lambda = 0$ where we ssume that the parameters $\rho_2, \dots, \rho_k$ are chosen so that no root of θ_k is real. hus, the corresponding potentials are complex and the operator L_k not selfadjoint.

If one set

$$\Phi_j = W_k^{-1} W_k\left(\frac{x^j}{j!}\right) \quad \text{for} \quad j \geq 0$$

nd $\Phi_j = 0$ for $j = -1, -2$, one finds that

$$L_k \Phi_j + \Phi_{j-2} = 0 \quad \text{for} \quad j = 0,1,2,\dots k .$$

ut only the Φ_j for even j are linearly independent while those with odd sub- cript satisfy

$$\Phi_{2j-1} + \sum_{i=0}^{j-2} \rho_{j-i} \Phi_{2i} = 0 \quad \text{for} \quad j = 1,2,\dots k .$$

The results of this last section were prompted by discussions with .A. Marčenko to whom we express our thanks. The formulae of the present paper simplify greatly the discussion of rational solutions of the Korteweg-de Vries equation in [1].

References

[1] Airault, H., McKean, H.P. and Moser, J., Rational and Elliptic Solutions of the Korteweg-de Vries Equation and a Related Many Body Problem. Comm. Pure Appl. Math., Vol. 30, 1977.

[2] Crum, M.M., Associated Sturm-Liouville Systems. Quart. Journ. Math., Ser. 2, Vol. 6, 1955, 121-127.

[3] Dubrovin, B.A. and Novikov, S.P., A periodic problem for the Korteweg-de Vrie and Sturm-Liouville equations. Sov. Math. Dokl., Vol. 15, 1974, 1597-1601.

[4] Gardner, C.S., Greene, J.M., Kruskal, M.D. and Miura, R.M., Korteweg-de Vries Equation and Generalizations VI, Methods for exact solution. Comm. Pure Appl. Math., Vol. 27, 1974, 97-133.

[5] Its, A.R. and Matveev, V.B., Hill's operator with finitely many gaps. Funct. Anal. and its Appl., Vol. 9, No 1, 1975, 69-70.

[6] McKean, H.P. and Trubowitz, E., Hill's Operator and Hyperelliptic Function theory in the presence of infinitely many branchpoints. Comm. Pure Appl. Math., Vol. 29, 1976, 143-226.

[7] Moses, H.E. and Tuan, S.F., Potentials with Zero Scattering Phase. Il Nuovo Cimento, Ser. 10, Vol. 13, 1959, 197-206.

Differential games of evasion

Czesław Olech (WARSAW)

Introduction. Consider a set valued measurable mapping Q from $[0,1]$ into closed subsets R^n and denote by A_Q the set of integrable selections of Q; that is the set of all $v : [0,1] \to R^n$ integrable and such that $v(t) \in Q(t)$ almost everywhere (a.e.) in $[0,1]$. It is known that L_1-weak closure of A_Q equals $A_{\text{clco } Q}$, where clco stands for closed convex hull. This implies that for each $\varepsilon > 0$ there is a map $\Phi_\varepsilon : A_{\text{clco } Q} \to A_Q$ such that for each $v \in A_{\text{clco } Q}$

$$(1) \qquad \left|\int_0^t (v(s) - \Phi(v)(s))ds\right| \leq \varepsilon , \quad t \in [0,1]$$

The aim of this paper is to prove that Φ_ε may be chosen such that besides (1) it satisfies

(2) if $v_1(s) = v_2(s)$ for $0 \leq s \leq T$ then $\Phi(v_1)(s) = \Phi(v_2)(s)$ for $0 \leq s \leq T$.

This is contained in a lemma due to B.Kaśkosz [7]. The version presented here is slightly more general and the proof differs from the original one of Kaśkosz by making the role of differential inequalities more explicit.

The need for such construction comes from differential games of evasion and it is instrumental in establishing the fact that in some cases mixed strategies of evasion can be approximated by ordinary ones. These connections with evasion problem are explained in detail in the next section. Section 3 contains the precise statement of the lemma and its proof, while Section 4 contains the proof of the approximation theorem stated in section 2.

2. Evasion problem.

A differential game is described by the system

$$(3) \qquad \dot{z} = P(z,u,v), \ z \in R^n, \ u \in U \subset R^m, \ v \in V \subset R^k$$

with two control parameters controlled by different players. There is a set M given and the aim of one player, say u, is to control the system in such a way that the resulting trajectory hits M in some finite time while the other tries to avoid M. The first is called the pursuer and the second - the evader. Both control the system by choosi their control functions from a class of admissible controls. In our ca admissible controls are A_U for the pursuer, and A_V for the evader that is all measurable functions taking values from fixed sets U and respectively. It is natural to assume that the choice depends on the past behaviour of the opposer but not of the future, since the future is not known for both.

Thus the following definition is motivated.

By a strategy of the evader one means a mapping $\psi : A_u \to A_v$ which satisfies condition (2). This expresses the fact that the evader has no knowledge of the future decisions of the pursuer. This i an ordinary strategy. The evasion problem has a positive solution if the is a strategy ψ for the evader such that the solution of the Cauchy problem

$$\dot{z}(t) = P(z(t), u(t), \psi(u)(t)) \ , \ z(0) = z_o$$

for any admissible $u \in A_U$ satisfies the condition

$$z(t) \notin M \quad \text{for each } t > 0.$$

Clearly, for existence of such strategies, which make the evas possible, one needs an assumption that the evader can influence the sys more than the pursuer.

Such conditions have been found and we should mention papers by S.Pontrjagin [1], [2], E.F.Mishchenko [3], R.V.Gamkrelidze and G.L. haratishvili [4], M.S.Nikolski [5] and P.B.Gusatnikov [6].Also the paper Kaśkosz [7] gives sufficient conditions for the existence of a solution of e evasion problem.

In these papers M is assumed to be a linear subspace of co- mension at least two. The following is perhaps the most simple version the evasion condition in question.

For each $z_* \in M$ there is a two dimensional subspace L ortho- nal to M such that

.) $$\text{int} \bigcap_{u \in U} \pi P(z_*, u, V) \neq \emptyset,$$

iere π stands for the orthogonal projection on L.

Suppose now that the evader, instead of choosing a value from V, efines for each t a probability distribution v_t on V. In this case ie response is given by

5) $$\dot{z}(t) = \int_V P(z(t), u(t), v)\, dv_t(v) ,$$

ence $$\dot{z}(t) \in \text{clco } P(z(t), u(t), V)$$

ince the spaces in question are of finite dimension, this can be xpressed in the following way

5) $$\dot{z}(t) = \sum_{i=o}^{n} \mu_i(t)\, P(t, u(t), v_i(t))$$

iere μ_i, v_i are measurable $\mu_i \geq 0$, $\sum_{i=o}^{n} \mu_i = 1$ and $v_i \in A_V$. Thus the vader can restrict himself without any loss of generality to probabilit istributions centered in n + 1 points of V. These remarks lead to ie definition: A mixed strategy is a map $\tilde{\psi}$ on A_U, whose values are $\mu_o, \ldots, \mu_n, v_o, \ldots, v_n), \mu_i, v_i$ measurable, $\mu_i \geq 0$, $\Sigma\mu_i = 1$ and $v_i \in A_V$ iich satisfy condition (2).

The class of mixed strategies is larger than that of ordinary ones. In particular condition (4), if mixed strategies are concerne[d] can be relaxed to

$$(4') \qquad \text{int} \bigcap_{u \in U} \text{clco}\, \pi\, P(z_*,u,V) \neq \emptyset$$

The result mentioned in the introduction can be used to prove that (4') is also sufficient for the existence of an ordinary strategy of evasion or more general that mixed strategies can be replaced (approximated) by ordinary ones. This follows from the following theorem due to Kaśkosz [7].

Theorem. Assume that $P(z,u,v)$ is continuous, satisfies a Lipschi[tz] condition with respect to z with a constant K, and that $P(0,u,v)$ is uniformly bounded. Then for each $\eta > 0$ and $T > 0$ and any mixed strategy $\tilde{\psi}$ there is an ordinary strategy ψ_η such that for any cont[inuous] function $u \in A_U$ the solution z_1 of (3) corresponding to u and $\psi_\eta u$, and the solution z_2 of (6) corresponding to u and $\tilde{\psi}u$ with the same fixed initial condition z_o, satisfy the inequality

$$|z_1(t) - z_2(t)| \leq \eta \quad \text{for} \quad 0 \leq t \leq T.$$

The proof of the theorem will be given in Section 4.

3. Kaśkosz' Lemma.

The result we mentioned in the introduction is a special case of the theorem which we shall now prove.

Lemma. Let $Q(t)$ be a set-valued map from $[0,1]$ into closed subsets of R^n and assume that Q is measurable and integrably bounded by f; i.e. that for each measurable $v : [0,1] \to R^n$, $v(t) \in Q(t)$ the inequality holds: $|v(t)| \leq f(t)$, where f is integrable on $[0,1]$.

Then for each $\varepsilon > 0$ there is a map

$$\Phi : A_{clco\,Q} \to A_{\Omega}$$

atisfying (1) and (2).

Proof. Let us fix $v \in A_{clco\,Q}$ ($v(t) \in clco\Omega(t)$) a.e. in $[0,1]$)
nd put $x(t) = \int_0^t v(s)ds$. Define the set-valued map $S(t,w)$ by the
ormula

(7) $$S(t,w) = \{q \in Q(t) \mid (q - v(t))(w - x(t)) \leq 0\} ,$$

here $(\cdot)(\cdot)$ stands for the scalar product.

Notice that $S(t,w) \subset Q(t)$ and that it is upper semicontinuous
ith respect to w for fixed t a.e. in $[0,1]$.

Consider the differential inclusion

(8) $$\dot{w} \in S(t,w)$$

Notice that because of (7) any solution $w(t)$ of (8) has the
roperty that $|w(t) - x(t)|$ is not increasing. In particular, if (8)
ad a solution with initial condition $w(0) = 0$, then it would be unique
nd equal to $x(t)$. But $S(t,x(t)) = Q(t)$ and $v(t)$ does not belong to
$Q(t)$, thus (8) does not admit a solution with initial condition
$w(0) = 0$.

Consider an approximate solution w_δ of (8) defined as follows

(9) $$\dot{w}_\delta(t) \in S(t,w_\delta(t_n)),\ t_n \leq t \leq t_{n+1},\ w_\delta(o) = 0,$$

where w_δ is continuous and $t_n = n\delta$, $\delta > o$.

In the formula

$$\frac{d}{dt}(w_\delta(t) - x(t))^2 = 2(\dot{w}_\delta(t) - v(t))(w_\delta(t_n) - x(t)) +$$

$$+ 2(w_\delta(t) - v(t))(w_\delta(t) - w_\delta(t_n)),\quad t_n \leq t \leq t_{n+1}$$

we see by (7) and (9) that the first part of the right-hand side

is non-positive, while the second can be estimated by $2f(t) \max \int_t^{t+\delta} f$
Thus

(10) $$|w_\delta(t) - x(t)| \leq k(\delta) \ , \ 0 \leq t \leq s,$$

where $k(\delta)$ tends to zero as $\delta \to 0$.

Notice that the estimate (10) does not depend on the choice o
so that if $v_1(t) = v_2(t) \quad 0 \leq t \leq \tau, \ n\delta \leq \tau < (n+1)\delta$, then the right hand side of (8) corresponding to v_1 or v_2 is the same for $t \leq \tau$. To complete the proof we therefore need to define a solution of (
in a unique way. This can, for example be done by putting $\dot{w}_\delta(t)$ as the lexicographical maximum of $S(w_\delta(t_n),t)$.

4. Proof of the approximation theorem.

Because of the Lipschitz condition all solutions with the fixed initial condition of (3) are uniformly bounded. Therefore $Q(t) = P(z(t),u(t),V)$ is integrably bounded and the bounding function f ca be made independent of the solution $z(t)$ of (3) and of the admissible co function u. Now let us fix u, and let $z_2(t)$ be the solution of (6) corresponding $\{\mu_i, v_i\} = \tilde{\psi}u$. Then $\dot{z}_2(t) \in \text{clco}\, P(z_2(t), u(t),V)$. Put $Q(t) = P(z_2(t), u(t),V)$. By the Lemma there is a $w(t) \in Q(t)$ such that

(11) $$|z_2(t) - z_o - \int_o^t w(s)ds| < \varepsilon,$$

and the map $u \to w$ is independent of the future. Let $v \in A_V$ be such that $w(t) = P(z_2(t), u(t), v(t))$. Again using lexicographical order we can make v uniquely defined. We put $v = \psi_{\eta(\varepsilon)}u$. We note that the defined map $\psi_{\eta(\varepsilon)} : A_U \to A_V$ also satisfies (2).

Now, put

$$\dot{z}_1(t) = P(z_1(t),u(t), (\psi_{\eta(\varepsilon)}u)(t)), \ z_1(0) = z_o$$

and let us estimate the difference $|z_1(t) - z_2(t)|$.

e have

$$|z_2(t) - z_1(t)| \leq | z_2(t) - z_o - \int_o^t P(z_2(t),u(t),v(t))dt| +$$

$$+ \int_o^t |P(z_2(t),u(t),v(t)) - P(z_1(t),u(t),v(t))|dt$$

Hence by (11) and the Lipschitz condition we get an integral nequality

$$|z_2(t) - z_1(t)| \leq \varepsilon + K \int_o^t |z_2(t) - z_1(t)|dt ,$$

hich implies that $|z_2(t) - z_1(t)|$ can be estimated on $[0,1]$ by a onstant $\eta(\varepsilon)$ which tends to zero as $\varepsilon \to 0$. This completes the proof.

Remark. The above proof is the same as the original proof of he theorem given by Kaśkosz (see [7]).

References

[1] L.S.Pontrjagin, A linear differential game of escape (Russian), Trudy Mat.Inst.Steklov. 112 (1971), pp. 30-63

[2] L.S.Pontrjagin, E.F.Mishchenko, On a problem of avoidance in linear differential games (Russian), Differencial'nye Uravnenija 7 (1971), pp. 436-445

[3] E.F.Mishchenko, On the problem of evading the encounter in differential games, SIAM J.Control 12 (1974), pp.300-310

[4] R.V.Gamkrelidze, G.L.Kcharatishvili, A Differential game of evasion with nonlinear control, SIAM J.Control 12 (1974), pp. 332-349

[5] M.S.Nikolski, On a quasi-linear problem of evading (Russian), Dokl.Akad.Nauk SSSR 221 (1975), pp.539-543

[6] P.B.Gusatnikov, On a problem of l-evading (Russian), Prikl.Mat. Meh. 40(1976) , pp.25-47

[7] B.Kaśkosz, On a nonlinear evasion problem, to appear in SIAM J. on Control and Optimization.

DIFFERENCE EQUATIONS AND WEYL'S LIMIT-POINT AND LIMIT-CIRCLE THEORY

by

Åke Pleijel.

Let $I = \{x_s\}_{-\infty}^{\infty}$, $s = \dots -2, -1, 0, 1, 2, \dots$, be a set of increasing real numbers. A difference equation on this set is

$$(Pu)(x_s) = f(x_s) \quad \text{with} \quad (Pu)(x_s) = \sum_{i=\alpha}^{\beta} p_i(x_s)u(x_{s+i}), \quad \alpha \text{ and } \beta \text{ integers.}$$

The coefficients p_i , $i = \alpha, \alpha+1, \dots \beta$, are complex valued functions on I , f is a given function of the same kind, and u shall satisfy the equation for all x_s. The equation is regular of the order $\beta - \alpha$ if $p_\alpha(x_s) \neq 0$, $p_\beta(x_s) \neq 0$ for all x_s. The points x_s can be, and often are, denoted by their indices only. F.V. Atkinson, [1] , chapter 5, has generalized Weyl's classical limit-point and limit-circle theory for differential equations to formally symmetric second order difference equations $Pu = \lambda u$ with real coefficients on a "half-infinite" interval $I = \{x_s\}_{s=0}^{\infty}$. C. Billigheimer, [2] , has extended Atkinson's theory to fourth order equations by a method which works also for equations of higher even orders. The author, [3] , has extended Weyl's theory to ordinary differential equations of the type $Su = \lambda Tu$, where S and T are formally symmetric differential operators one of which has a positive Dirichlet integral. As a generalization the method includes Weyl's original theory and its subsequent extensions to higher order differential equations.

The purpose of the present paper is to indicate how the method applies to difference equations. To avoid technicalities only equations of the type $Su = \lambda u$ are considered on a double infinite interval. The method can, however, be used also for equations $Su = \lambda Tu$, where S and T are formally symmetric difference operators one of which has a positive character. The treatment of a finite or half infinite interval instead of a double infinite requires certain modifications.

To parallel the treatment with the one used for differential equations, an integral of a function u is introduced as the additive set function which for very finite interval J (containing all x_s between included or excluded endpoints x_n and x_m) is defined by

$$\int_J u = \sum_{x_s \in J} u(x_s).$$

asic is a Lagrange identity which depends upon "partial integration". If τ is the ranslation operator $(\tau u)(x_s) = u(x_{s+1})$, it is easily seen that

$$\int_J \tau u = \int_J u + {}_J[u]$$

ith ${}_J[u] = u(\tau J - J) - u(J - \tau J)$. Here τJ is the interval of all points x_{s+1} such that x_s belongs to J. The definition of ${}_J[u]$ is appropriate since $\tau J - J$ and $J - \tau J$ each contains only one point being either an endpoint of J or its translation one step to the right. Because of $\tau(uv) = \tau u \cdot \tau v$ it follows that

$$\int_J \tau u \cdot \overline{v} = \int_J u \cdot \overline{\tau^{-1} v} + {}_J[u \cdot \overline{\tau^{-1} v}], \tag{1}$$

and that

$$\int_J \tau^{-1} u \cdot \overline{v} = \int_J u \cdot \overline{\tau v} - {}_J[\tau^{-1} u \cdot \overline{v}]. \tag{2}$$

Therefore the formal adjoints of τ and τ^{-1} are $(\tau)^* = \tau^{-1}$ and $(\tau^{-1})^* = \tau$. Similarly, the formal adjoint of the multiplication by a function a is the multiplication by its complex conjugate, $(a)^* = \overline{a}$. According to $(P+Q)^* = P^* + Q^*$ and $(PQ)^* = Q^* P^*$, the finite sum

$$S^* = \Sigma\, \tau^{-j}\, \overline{a}_{kj} \tau^k, \quad j \text{ and } k \geq 0,$$

is the formal adjoint of

$$S = \Sigma\, \tau^{-j} a_{jk} \tau^k, \text{ finite sum, } j \text{ and } k \geq 0. \tag{3}$$

The operator S is formally symmetric, $S^* = S$, if the coefficients a_{jk} are hermitean, $\overline{a}_{jk} = a_{kj}$. Because of $\tau^{-j} a_{jk} \tau^k = (\tau^{-j} a_{jk}) \cdot \tau^{k-j}$, and $\tau^{-j} a_{jk} \tau^k = \tau^{-j} \tau^k \tau^{-k} a_{jk} \tau_k = \tau^{-j+k} \cdot (\tau^{-k} a_{jk})$, such a formally symmetric operator S can be reduced to

$$S = \sum_{k=0}^{m} a_{0k} \tau^k + \sum_{k=1}^{m} \tau^{-k}\, \overline{a}_{0k} \tag{4}$$

where both sums end with the same integer $k = m$. This is the form of operators considered by Atkinson and Billigheimer (with real coefficients). It is supposed that

$a_{0m}(x_s) \neq 0$ for all x_s which implies that S is regular and has order $M = 2m$.

For a regular operator S of the order M, a solution of $(Su)(x_s) = f(x_s)$ for a given f is uniquely determined by the values of u on a set of M consecutive points. The homogeneous equation $(Su)(x_s) = 0$ therefore has M and no more linearly independent solutions $u_1, \dots u_M$, and any solution of the inhomogeneous equation can be written $u = w + c_1u_1 + \dots + c_Mu_M$, where w is a particular solution and $c_1, \dots c_M$ are constants.

For an operator represented as in (3), a partial integration formula is obtained by iterated use of (2), namely

$$\int_J Su \cdot \overline{v} = {}_J(u,v)^S - \Big[\sum_{\rho=1}^{m} S^\rho u \cdot \overline{\tau^{\rho-1}v}\Big]_J, \tag{5}$$

where

$${}_J(u,v)^S = \int_J \sum a_{jk}\tau^k u \cdot \overline{\tau^j v} \tag{6}$$

and

$$S^\rho = \tau^{\rho-1} \sum_{\substack{j \geq \rho \\ k \geq 0}} \tau^{-j} a_{jk} \tau^k .$$

Under the assumption $\overline{a}_{jk} = a_{kj}$, the form ${}_J(u,v)^S$ is hermitean, i.e. ${}_J\overline{(u,v)}^S = {}_J(v,u)^S$, and ${}_J(u,v)^S$ is linear in u. By the use of (4) instead of (3) the formula (5) is not structurally simplified. ${}_J(u,v)^S$ being hermitean, a consequence of (5) is the Lagrange formula

$$\int_J Su \cdot \overline{v} - \int_J u \cdot \overline{Sv} = - \Big[\sum_{\rho=1}^{m} S^\rho u \cdot \overline{\tau^{\rho-1}v} - \tau^{\rho-1}u \cdot \overline{S^\rho v}\Big]_J .$$

With

$${}_JQ(u,v) = i^{-1}\Big(\int_J Su \cdot \overline{v} - \int_J u \cdot \overline{Sv}\Big), \tag{7}$$

and

$$q(u,v) = \sum_{\rho=1}^{m} \big(iS^\rho u \cdot \overline{\tau^{\rho-1}v} + \tau^{\rho-1}u \cdot \overline{iS^\rho v}\big),$$

the formula can be written

$${}_JQ(u,v) = \big[q(u,v)\big]_J .$$

Both ${}_JQ(u,v)$ and $q(u,v)$ are hermitean. For $v = u$

$$q(u,u) = \frac{1}{2}\sum_{\rho=1}^{m} |iS^{\rho}u + \tau^{\rho-1}u|^2 - \frac{1}{2}\sum_{\rho=1}^{m} |iS^{\rho}u - \tau^{\rho-1}u|^2 .$$

This identity shows that on any finite dimensional space (spanned by a finite number of functions u on I), the signature i.e. the pair of positive and negative inertia indices of q satisfies the inequality

$$\text{sig } q \leq (m,m) .$$

as a consequence, and in a similar meaning,

$$\text{sig } {}_{J}Q \leq (M,M) , \tag{8}$$

where $M = 2m$. There exist finite dimensional spaces on which the maximum (M,M) is attained. For, let E_λ be the space of all solutions of $Su = \lambda u$ for a non-real value of λ. E_λ has the dimension M (order of the difference equation), and E_λ and $E_{\bar{\lambda}}$ have only the nullelement in common so that the direct sum $F_\lambda = E_\lambda \dotplus E_{\bar{\lambda}}$ exists and is $2M$-dimensional. For u and v in E_λ, (7) can be reduced to

$${}_{J}Q(u,v) = c\ {}_{J}(u,v) ,$$

where $c = c(\lambda) = i^{-1}(\lambda - \bar{\lambda})$, and the right hand side contains the hermitean form

$${}_{J}(u,v) = \int_{J} u \cdot \bar{v} = \sum_{x_s \in J} u(x_s)\overline{v(x_s)} . \tag{9}$$

Since ${}_{J}(u,u) = \sum_{x_s \in J} |u(x_s)|^2$, the form $c\ {}_{J}Q$ is positive on E_λ, and definite provided J contains at least M points. With the same value $c = c(\lambda)$, it is negative definite on $E_{\bar{\lambda}}$, hence has the signature (M,M) on $F_\lambda = E_\lambda \dotplus E_{\bar{\lambda}}$. The form ${}_{J}Q$ has the same signature.

Weyl's theory is a study of ${}_{J}Q$. Let J and J' be finite intervals such that $J \subset J'$. According to (7)

$${}_{J'}Q(u,u) - {}_{J}Q(u,u) = i^{-1}\left(\int_{J'-J} Su \cdot \bar{u} - \int_{J'-J} u \cdot \overline{Su}\right) .$$

Put $Su = \lambda u + f$, where λ is non-real. Then

$$c\ {}_{J'}Q - c\ {}_{J}Q = c^2\ {}_{J'-J}(u,u) - ic\ {}_{J'-J}(f,u) + ic\ {}_{J'-J}(u,f) , \quad c = i^{-1}(\lambda - \bar{\lambda}) .$$

By the help of Cauchy-Schwarz inequality for the scalar product (9) integrated over $J'-J$, the following inequality is obtained

$$(c\ {}_{J'}Q(u,u) - {}_{I_0-J'}(f,f)) - (c\ {}_{J}Q(u,u) - {}_{I_0-J}(f,f)) \geq (|c|\ {}_{J'-J}|u| - {}_{J'-J}|f|)^2$$

where ${}_{J'-J}|u| = \sqrt{{}_{J'-J}(u,u)}$, same for f, and I_0 is a finite interval containing J' and J. This holds true provided $Su = \lambda u + f$, hence also if u is replaced by $u-v$, where v belongs to E_λ. If

$$ {}_{I}(f,f) = \int_I |f|^2 = \sum_{x_s \in I} |f(x_s)|^2 < \infty, \tag{10}$$

I_0 may be replaced by the entire interval I, and

$$\begin{aligned}(c\ {}_{J'}Q(u-v,u-v) - {}_{I-J'}(f,f)) - (c\ {}_{J}Q(u-v,u-v) - {}_{I-J}(f,f)) \geq \\ \geq (|c|_{J'-J}|u-v| - {}_{J'-J}|f|)^2 .\end{aligned} \tag{11}$$

For any f, the equation $Su - \lambda u = f$ has a solution u, and if f satisfies (10) the inequality (11) holds true for all v in E_λ with a non-real value of λ. In connection with (11) analogues of Weyl's circles are defined as the sets

$$\Sigma\ {}_{J}(u) = \{v \in E_\lambda : c\ {}_{J}Q(u-v,u-v) - {}_{I-J}(f,f) \leq 0\}$$

in the M-dimensional space E_λ. Such a set is bounded by a second order surface in the variable point v of E_λ ($v = t_1u_1 + \ldots + t_Mu_M$, where $u_1, \ldots u_M$ are linearly independent solutions of $Su = \lambda u$). Since $c\ {}_{J}Q$ is positive definite on E_λ, the boundary is of ellipsoid type. Because of (11) the sets are nested, shrinking when J is increased. They are non-empty as can be seen by considering $v = u(J)$ such that

$$c\ {}_{J}Q(u - u(J),\ v) = 0,\ u(J) \in E_\lambda,$$

for all v in E_λ. This $u(J)$ is uniquely determined since $c\ {}_{J}Q$ is positive definite and therefore non-degenerate on E_λ. It is asserted that

$$c\ {}_{J}Q(u - u(J),\ u - u(J)) \leq 0 \tag{12}$$

which proves that $u(J)$ belongs to $\Sigma\ {}_{J}(u)$. The assumption contrary to (12) would imply that $c\ {}_{J}Q$ is positive definite on the $(M+1)$-dimensional linear hull $\{u - u(J), E_\lambda\}$ which is contradicted by (8), $\operatorname{sig} c(\lambda)\ {}_{J}Q \leq (M,M)$.

Being compact, all ellipsoids $\Sigma\ {}_{J}(u)$ contain at least one point v for which according to (11) the expressions $c\ {}_{J}Q(u-v,u-v) - {}_{I-J}(f,f)$ tend increasingly to a limit when J tends to I. Referring again to (11) a consequence is that its left hand side, then also the right hand side tends to 0 when J and J' tend to I. During this transition to the limit, ${}_{J'-J}|f|$ tends to 0 since ${}_{I}(f,f)$ exists. Therefore the limit of ${}_{J'-J}|u-v|$ is also 0 which proves that

$$ {}_{I}(u-v,u-v) < \infty, $$

where $Su - \lambda u = f$, f satisfies (10) and v is an element in E_λ, i.e. $Sv - \lambda v = 0$. This means that for any f satisfying

$$ \sum_{x_s \in I} |f(x_s)|^2 < \infty $$

the difference equation $Su - \lambda u = f$ has at least one solution u such that

$$ \sum_{x_s \in I} |u(x_s)|^2 < \infty. $$

This result corresponds to a fundamental statement in Weyl's theory concerning the existence of square integrable solutions of differential equations. Due to elementary properties of hermitean forms, a continued study of ${}_{J}Q$ as in [3] shows that the number of linearly independent solutions of $Su - \lambda u = 0$ such that

$$ {}_{I}(u,u) = \sum_{x_s \in I} |u(x_s)|^2 < \infty \tag{13} $$

is the same, p, for all λ in $\mathrm{Im}(\lambda) > 0$ and the same, n, for all λ in $\mathrm{Im}(\lambda) < 0$. Furthermore, in case $p = n$, selfadjoint restrictions of S exist in the hilbertspace of functions u for which (13) holds true, such restrictions being characterized by symmetric boundary conditions.

Equations

$$ (\Sigma\, \tau^{-j} a_{jk} \tau^{k} u)(x_s) = \lambda\, b(x_s) u(x_s) $$

can be treated as indicated above if $b(x_s) > 0$ for all x_s, by introducing

$$ {}_{J}(u,v) = \int_J buv = \sum_{x_s \in J} b(x_s) u(x_s) \overline{v(x_s)} \tag{14} $$

which is then positive definite. If the positivity of b is not fulfilled, the representation of S should not be reduced to the form (4). If possible, a representation (3) should be used such that the hermitean form

$$_J(u,v)^S = \int_J \Sigma\, a_{jk} \tau^k u \cdot \overline{\tau^j v}\,, \tag{15}$$

has the property $_J(u,u)^S > 0$ for $u \neq 0$. The form (15) can then be used as a scalar product instead of the non-definite form (14). A second order operator S has for instance the needed positivity if

$$a_{11}(x_s) > 0\,, \quad \begin{vmatrix} a_{00}(x_s) & a_{01}(x_s) \\ a_{10}(x_s) & a_{11}(x_s) \end{vmatrix} > 0$$

for all x_s. For a treatment of $Su = \lambda Tu$ the use of suitable representations of the operators S and T is also needed.

REFERENCES.

[1] Atkinson, F.V., Discrete and Continuous Boundary Problems. Academic Press, New York and London 1964.

[2] Billigheimer, C.E., Singular boundary problems for a five-term recurrence relation. Amer. J. Math. Vol. XCI, No. 4 (1969), 1012-1048.

[3] Pleijel, Åke, A survey of spectral theory for pairs of ordinary differential operators. Lecture Notes in Mathematics 448 (1975), 256-272.

ADDITIONAL NOTE. The author regrets that he was unaware of A. Schneider's treatment of left-definite difference systems extending Billigheimer's results and published in J. Reine Angew. Math., Band 280, Seite 70-76 und 77-90.

Dept. of Math.,
Sysslomansgatan 8,
S - 752 23 Uppsala,
Sweden.

SQUARE-INTEGRABLE SOLUTIONS, SELF-ADJOINT EXTENSIONS AND SPECTRUM OF DIFFERENTIAL SYSTEMS

F.S. Rofe-Beketov

The recent results obtained by the author and his co-workers are surveyed. The proofs for some author's results are outlined. The bibliography can be found in the papers cited.

1. Different forms of representation of self-adjoint and dissipative relations. (See [5],[6]). In the theory of extensions of operators the self-adjoint [1] and dissipative [2],[3] relations in the Hilbert space of boundary values are of great importance. A thorough survey of the theory of relations, produced by pairs of differential operators, is given in [4] . Describing the self-adjoint boundary conditions for differential equations with bounded operator coefficients [5],[6] , the author (unfortunately, knowing nothing about the paper [1]) reintroduced the concept of a self-adjoint relation (having called it Hermitian) and obtained canonical forms for such relations: $\cos A\cdot x' - \sin A\cdot x = 0$, where $x, x' \in H$, A is a bounded self-adjoint operator or

$$(U - I)\, x' + i\, (U + I)\, x = 0, \tag{1.1}$$

where U is the unitary operator. But if U is the contraction, the relation (1.1) is maximally dissipative (see [2], [3]). Relations are not however found in the canonical form at all times.

Theorem 1.1. (Cf. [5] ,[6]).Let B and C be arbitrary bounded operators defined everywhere in H. For the relation θ to be maximally dissipative:

$$x\,\theta\, x' \Longleftrightarrow Cx' - Bx = 0 \tag{1.2}$$

it is necessary and sufficient that the operator $B+iC$ be injective and the operator

$$U = (B + iC)^{-1}\,(B - iC) \tag{1.3}$$

be the contraction (defined everywhere in H). Under these conditions U leads to the canonical form (1.1) of the relation θ and is its Kelly transform.

Proof. Necessity. If (1.2) is maximally dissipative, it follows from the representation of θ in the form (1.1) that $\exists\, v \in H$: $x=(U-I)v$, $x' = -i(U+I)v$, i.e. $-iC(U+I) - B(U-I) = 0$, from which $C=K(U-I)$, $B= -Ki(U+I)$ for a certain injective operator K. But then $B+iC= -2iK$ is also injective, $B-iC = -2iKU$, and there is $(B+iC)^{-1}(B-iC) = U$, where U is the contraction.

Sufficiency. Let (1.3) be the contraction. Substituting (1.3) into (1.1), we can find (1.2).

For (1.2) to be self-adjoint, the unitarity of U (1.3) is necessary and sufficient (see [5],[6]). The case when B and C of (1.2) can be unbounded, unclosed and have nondense-in-H domains was also considered there.

2. Differential systems with the Nevanlinna's weight function (see [7]-[9]). Many dissipative differential equations and systems of an arbitrary order are reduced to the system:

$$\frac{1}{2} i \left\{ \left[Q(t)y \right]' + Q^*(t)y' \right\} = \left[A(t,\lambda) + B(t) \right] y, \qquad (2.1)$$

where A, B, Q are (n x n)- matrices, $\det \left[\operatorname{Re} Q(t) \right] \neq 0$, $Q(t) \in AC_{loc}$,

$$\operatorname{Im} A(t,\lambda) \geqslant 0 \quad \text{for } \operatorname{Im}\lambda > 0; \ \operatorname{Im} B(t) \geqslant 0 \qquad (2.2)$$

$A(t,\lambda) \in L^1_{loc}$ in t and is analytic in $\lambda \in C_+$, $B(t) \in L^1_{loc}$.

Call the A-deficiency number $N_A(\lambda)$ of the system (2.1) on a half-axis R_+ the maximal number of linearly independent solutions $y(t) \in E_n$ of (2.1) which belong to $L^2_{A,\lambda}(R_+)$, i.e.

$$\int_0^\infty y^*(t) \left[\operatorname{Im} A(t,\lambda) \right] y(t)\, dt < \infty \qquad (2.3)$$

The metrics $L^2_{A,\lambda}(R_+)$ are equivalent for $\lambda \in C_+$. The equality $\det[\operatorname{Im} A(t,\lambda)] = 0$ is not excluded. The case $A(t,\lambda) = \lambda A(t)$, $B^*(t) = B(t)$ is considered in [7],[8] by using methods of the theory of functions, the characteristic Weyl matrix, etc. In this case anothe approach [10] is based on the symmetric relation theory, but it presents some difficulties for investigation of the general problem (2.1).

<u>Lemma 2.1.</u> By the non-degenerate substitution $y(t,\lambda) = S(t)x(t,\lambda$ the system (2.1) can be reduced to the form:

$$J \frac{dx}{dt} = H(t,\lambda)x, \quad H(t,\lambda) = W(t,\lambda) + C(t) \quad (2.1c$$

where $J^2 = -I_n$, $J^* = -J = \text{Const}$, $\operatorname{Im} W(t,\lambda) \geqslant 0$ $(\operatorname{Im}\lambda > 0)$, $\operatorname{Im} C(t) \geqslant 0$ and the $L^2_{A,\lambda}(R_+)$- solutions of (2.1) correspond to the $L^2_{W,\lambda}(R_+)$-solutions of (2.1c) and vice versa. (Cf. Lemma 1.1 from [7]).

For systems of the form (2.1c) the constancy of the deficiency index $N_H(\lambda)$ $(\lambda \in C_+)$ for $C(t) \equiv 0$ was found by Orlov [11] who used different method based on studies of the Weyl matrix circles.

Theorem 2.1. [9]. For system (2.1),(2.2) on the half-axis $N_A(\lambda) = \text{Const}$ for $\lambda \in C_+$. ($N_A(\lambda)$ and $N_{A+B}(\lambda)$ can differ from one another).

Proof. It is sufficient to show that $N_W(\lambda) = \text{Const}$ ($\text{Im}\,\lambda > 0$) for (2.1c), assuming also that for non-trivial solutions of (2.1c)

$$\int_0^1 x^*(t,\lambda)\ \text{Im}\,W(t,\lambda)\ x(t,\lambda)\,dt > 0 \qquad (2.4)$$

(cf. Theorem 1.2 from [7]).

Lemma 2.2. For the non-homogeneous problem

$$J\frac{d\Psi}{dt} = H(t,\lambda)\Psi + f(t),\ t\in R_+,\quad \text{Im}\,\lambda>0, \qquad (2.5)$$

there is a resolvent kernel $K(t,\tau,\lambda)$ such that:

a) $$\int_0^\infty K(t,\tau,\lambda)\left[\text{Im}\,H(\tau,\lambda)\right] K^*(t,\tau,\lambda)\,d\tau < \infty \qquad (2.6)$$

b) if $f^*(t)\,l(t)\in L^1(R_+)$ for $\forall\, l(t)\in L^2_{W,\lambda}(R_+)\cap C$, in particular, if $|f(t)|\in L^1(R_+)$ and is finite, the vector-function

$$\Psi(t,\lambda) \overset{df}{=} \int_0^\infty K(t,\tau,\lambda)\,f(\tau)\,d\tau \qquad (2.7)$$

satisfies (2.5).

c) If in (2.7)

$$f(t) = \left[H(t,\lambda) - H(t,\mu)\right]g(t) = \left[W(t,\lambda) - W(t,\mu)\right]g(t), \qquad (2.8)$$

$\text{Im}\,\lambda>0$, $\text{Im}\,\mu>0$, $g(t)\in L^2_{W,\lambda}(R_+)$,

then

$$\int_0^\infty \Psi^*(t)\left[\text{Im}\,H(t,\lambda)\right]\Psi(t)\,dt \leq \frac{|\lambda-\mu|^2}{|\text{Im}\,\lambda|\cdot|\text{Im}\,\mu|}\int_0^\infty g^*(t)\left[\text{Im}\,W(t,\lambda)\right]g(t)\,dt \qquad (2.9)$$

The proof of the lemma is similar to that of lemma 2.1 from [7], provided the integral representation for Nevanlinna's functions being used. The proof of the theorem is completed in exactly the same manner as for Theorem 2.1 from [7], if we put $\Psi_p(t) = \int_0^\infty K(t,\tau,\mu)\left[H(\tau,\mu)-H(\tau,\lambda)\right]x_p(\tau,\lambda)\,d\tau$, where $x_p(t,\lambda)$ are linearliy independent $L^2_{W,\lambda}(R_+)$-solutions of the system (2.1c).

For the system (2.1) on the half-axis we put $N_A^+ = N_A(\lambda)$, ($\lambda\in C_+$), and similarly if

$$\text{Im}\,B(t)\leq 0,\quad A(t,\bar\lambda) = A^*(t,\lambda), \qquad (2.10)$$

then $N_A^- = N_A(\lambda)$, ($\lambda\in C_-$).

Theorem 2.2. The A-deficiency index of the syatem (2.1),(2.2) on the half-axis allows the estimate $\nu_-[\mathrm{Re}Q] \leqslant N_A^+ \leqslant n$, and for the condition (2.10) the estimate $\nu_+[\mathrm{Re}Q] \leqslant N_A^- \leqslant n$, where $\nu_\pm[\mathrm{Re}Q]$ is the number of positive (negative) eigenvalues of the matrix $\mathrm{Re}[Q(t)]$. (It is independent of t). (Cf.[7]). A particular case of this theorem is Theorem 1 from [12].

Let Im B (t)=0 in (2.1). If $N_A^+ = N_A^- = n$, the system (2.1) is called quasi-regular. The quasi-regularity is sometimes provided by a single condition $N_A^+ = n$ or $N_A^- = n$.

Theorem 2.3. If all the solutions of the system

$$\frac{1}{2} i\left\{[Q(t)y]' + Q^*(t)y'\right\} = A(t)y, \quad \mathrm{Im}A(t) \geqslant 0, \qquad (2.11)$$

belong to $L^2\{R_+, \mathrm{Im}A(t)dt\}$, then all the solutions of the system:

$$\frac{1}{2} i\left\{[Q(t)z]' + Q^*(t)z'\right\} = A^*(t)z \qquad (2.12)$$

belong to $L^2\{R_+, \mathrm{Im}\, A(t)\, dt\}$ if and only if

$$\inf_{0 \leqslant t < \infty} \int_0^t \mathrm{Tr}\left\{[\mathrm{Re}Q(\tau)]^{-1}\, \mathrm{Im}\, A(\tau)\right\} d\tau > -\infty \qquad (2.13)$$

The sufficiency of the condition (2.13) has been obtained in Theorem 3.1 from [7]. In fact the necessity has been also proved in [7](see[8]).Using Theorem 2.3,we can derive the result of [13].

Theorem 2.4. The A-deficiency indices $N_A(\lambda,R)$, $N_A(\lambda,R^\pm)$, of the system (2.1),(2.2) on the axis R and on the half-axes $R_\pm$ are related by

$$N_A(\lambda,R) = N_A(\lambda, R_-) + N_A(\lambda,R_+) - n, \quad (\lambda \in C_+), \qquad (2.14)$$

if for non-trivial solutions $y(t,\lambda)$ of the system (2.1) at some $\alpha<\beta$

$$\int_\alpha^\beta y^*(t,\lambda)\left[\mathrm{Im}\,(A(t,\lambda) + B(t))\right] y(t,\lambda)\, dt > 0 \qquad (2.15)$$

Without the condition (2.15), the equality (2.14) may no longer be true. The scheme of the proof is similar to those in [7] and [14]. Theorem 2.4 involves the result of [14].

It follows from (2.14) and Theorem 2.1, that if (2.15)holds,then

$$N_A(\lambda, R) = \mathrm{Const}, \quad (\lambda \in C_+). \qquad (2.16)$$

Theorem 2.5. If in (2.1) $A(t,\lambda)=\lambda A(t)$, $A(t)\geqslant 0$, $\mathrm{Im}\, B(t)\geqslant 0$ (or$\leqslant 0$), then (2.16) (or for $\lambda \in C_-$) is valid without the condition (2.15).

The proof follows from the fact that the system (2.1) gives rise in $L^2\{R, A(t)dt\}$ to an accumulative (or dissipative) relation,the deficiency index of which is constant for $\mathrm{Im}\lambda>0$ ($\mathrm{Im}\lambda<0$), and equal to $N_A(\lambda,R)$ minus the dimensionality of the null manifold of the system (2.1) which is constant too. (See Theorem 1.1 from [7]). For the case of symmetric system this proof was proposed by Bruk [10].

Equations of arbitrary order r=2 m or 2m + 1

$$L(\lambda)y=\sum_{k=0}^{m}(-1)^k(p_{m-k}y^{(k)})^{(k)}+\frac{1}{2}i\sum_{k=0}^{\left[\frac{r-1}{2}\right]}(-1)^k\left[(q_{m-k}y^{(k)})^{(k+1)}+\right.$$
$$\left.+(s_{m-k}y^{(k+1)})^{(k)}\right]=W(t,\lambda)\,y \qquad (2.17)$$

with the n x n- matrix coefficients $p_k(t,\lambda)$, $q_k(t,\lambda)$, $s_k(t,\lambda)$,$W(t,\lambda)$ can be reduced to the first-order system by using the quasi-derivatives introduced by formulae that are obtained by substituting $s_k(x,\lambda)$ for $q_k{}^*(x)$ in the quasi-derivatives from [7].

Theorem 2.6. Let the coefficients of (2.17) depend analytically on λ in a certain domain $D\subset C$ and let for any finite $\Delta\subset R_+$ and any smooth y (t)

$$\mathrm{Im}\,L_\Delta[y,y]\leqslant 0,\quad \mathrm{Im}\,W(t,\lambda)\geqslant 0\quad \text{for } \lambda\in D,\ t\in R_+,$$

where $L_\Delta[y,y]$ stands for the Dirichlet integral over the interval Δ , corresponding to the operation $L(\lambda)$ (2.17). Then:

a) if $W(t,\lambda)=W(t)$ is independent of λ, then $N_L(\lambda)$- number of such linearly independent solutions $y(t,\lambda)$ of the system (2.17), for which

$$\lim_{\Delta\to R_+}\left|\mathrm{Im}\,L_\Delta(\lambda)[y,y]\right|<\infty$$

is independet of $\lambda\in D$.

b) if the coefficients of operation $L(\lambda)$ are independent of λ, then the N_W-number of linearly independent solutions of the system (2.17), $y\in L^2_{W,\lambda}(R_+)$, does not depend on λ for $\lambda\in D$.

c) if r = 2m, then $mn\leqslant N_L(\lambda)$, $N_W(\lambda)\leqslant rn$ ($\lambda\in D$) and if r=2m+1, then $mn+\mathcal{V}_-[\mathrm{Re}q_0]\leqslant N_L(\lambda)$, $N_W(\lambda)\leqslant rn$, ($\lambda\in D$).

d) if $N_L(\lambda)=rn$ for the system $L(\lambda)y=0$ for some $\lambda\in D$, then for the adjoint system $L^*(\lambda)y=0$, ($t\in R_+$), $N_L(\lambda)=rn$ if and only if

$$\sup_t\int_0^t \mathrm{Im}\,\mathrm{Tr}\left[p_0^{-1}(q_0+s_0)\right]d\tau<\infty \quad \text{when } r=2m$$

or

$$\sup_t\int_0^t \mathrm{Tr}\left[(\mathrm{Re}q_0)^{-1}\,\mathrm{Im}\,p_0\right]d\tau<\infty \quad \text{when } r=2m+1$$

(if $r = 2m+1$ then $q_0 = s_0^*$ and are independent of λ).

Note that in the statements c) and d) a type of the dependence of the coefficients on λ is of no importance.

The proof can be obtained by reducing Eq. (2.17) to the first-order system. [9].

Note 1. Points b) and c) of Theorem 2.6 generalize Glazman's Theorem on dissipative operators ([15], Theorem 3).

Note 2. Putting $L(\lambda) = S-\lambda T$, where S,T are symmetric differential operators of arbitrary order (S or T having the Dirichlet non-negative integral over any finite $\Delta \subset R_+$), we can find that a number of results for the problem $Sy = \lambda Ty$ (see Survey [4]) follow for T-positive and S-positive cases simultaneously from the statements of Theorems 2.6 and 2.4, if the spectral parameter λ or $1/\lambda$ is a factor multiplying the operator that creates the metric.

An interesting example is the Sturm-Liouville equation with the sign-indefinite weight function, considered in the Dirichlet integral metric [16]. The m-coefficient of Weyl, derived for such an equation by Atkinson, Everitt and Ong [16], is relate to the Weyl characteristic matrix F for the first order system co responding to this equation by the formula:

$$F(\lambda) = \frac{1}{2} (\lambda m(\lambda)\cos\alpha + \sin\alpha)^{-1} \begin{bmatrix} 2\lambda\cos\alpha & \lambda m(\lambda)\cos\alpha - \sin\alpha \\ \lambda m(\lambda)\cos\alpha - \sin\alpha & -2m(\lambda)\sin\alpha \end{bmatrix}$$

Note also papers by Mikulina [17],[18] dealing with similar problems

3. On the discrete spectrum of the perturbed Hill's operator and of some differential systems. (See [19]-[24]).

Consider now perturbarions of the spectrum of Hill's operator under a combined perturbation of the potential q (x) both by a sma on infinity term p (x) and by a partial phase shift. Let

$$q_a(x) = \begin{cases} q(x), & x \geqslant 0, \\ q(x-a), & x<0, \quad a \in R, \end{cases} \tag{3.1}$$

$$L_a y \equiv -y'' + q_a(x)\, y = \lambda y, \tag{3.2}$$

Theorem 3.1. [19]. The essential spectrum $C(L_{a+p})$ of the problem

$$(L_a + p)y = -y'' + [q_a(x) + p(x)]y = \lambda y, \quad -\infty<x<\infty \tag{3.3}$$

provided that

$$p(x) \in L^1(-\infty, \infty) \tag{3.4}$$

coincides with $C(L_0)$ and bears no eigenvalues with the possible exception of the end-points of spectral lacunas, and for

$$(1 + |x|)\, p(x) \in L^1(-\infty, \infty) \qquad (3.5)$$

is entirely free of eigenvalues of (3.3). The theorem holds also for the equation on the half-axis. The functions q and p are complex-valued.

Theorem 3.2. [19]. Let q (x), p (x) be real-valued.

a) under the condition (3.5), every lacuna in $C(L_a+p)=C(L_o)$ contains not more than a finite number of eigenvalues of the problem (3.3), and each of sufficiently remote lacunas contains not more than two eigenvalues.

b) the statement a) holds for (3.3) on either of half-axis $[b,\infty)$ or $(-\infty,b]$ with the boundary conditions

$$\cos\beta \cdot y'(b) - \sin\beta \cdot y(b) = 0, \qquad (3.6)$$

each of the sufficiently remote lacunas containing not more than one eigenvalue.

A modification of the method of phase functions in the scattering theory has been proposed in [20], that enables it to be used in studying the eigenvalues within internal lacunas of the continuous spectrum rather than on the left-hand of C(L) only. E.g. for the perturbed Hill's operator the next theorem is true.

Theorem 3.3. ([19],[20]). Let $p(x) = p_1(x)+p_2(x)$ where $p_1(x)\in L^1(R)$ $p_2(x)$ is continuous and $\to 0$ for $|x|\to\infty$. If $p(x) \geqslant C(x^2+1)^{-1}$ then for any spectral lacuna of L_o (3.2) we can take such $C > 0$, that a left end-point of this lacuna will be a limiting point for the eigenvalues of the $(L_a + p)$ (3.3) which are located in the lacuna. But if $p(x) \leqslant C(x^2+1)^{-1}$, then for any $C>0$ the left end-points of sufficiently far lacunas will not be limiting points for the eigenvalues lying in these lacunas. The right end-points of lacunas are considered similarly. Exact values of the constants mentioned in the theorem can be given for every specific lacuna.

For self-adjoint periodic differential systems of any order, Hrabustovsky [21],[22] found sufficient conditions for the finiteness of the number of discrete levels that appear near a given end-point of spectral lacuna under symmetric perturbation affecting of all the coefficients. These conditions relate the rate of decreasing of the perturbation coefficients to the Jordan structure of monodromy matrix at a corresponding end-point of the lacuna.

The generalization of the Sturm oscillation theorem was given in [23] for the cases of finite and infinite systems of differential equations in the finite or infinite interval; it contains the Mors index-theorem as a particular case.

The conditions for discreteness of the spectrum of elliptic non symmetric systems with the coefficients of arbitrary rate of incre se at infinity were studied by Brusentsev [24]. Developing the methods used in [25], he showed that the Grushin-type determinant condition [26] for the operator symbol provides the discreteness of the spectrum of such systems without any requirement of the uniform ellipticity. Uniformly elliptic systems with coefficients of power rate of increase are considered in [26].

4. Consider now <u>the method of operator inequalities in the problems of essential self-adjointness of elliptic operators.</u> It has been shown ([27] - [30]), that for the essential self-adjointness of the Schrödinger Operator M and its powers and polynomials (when the potential is smooth), it is sufficient that the operator inequality

$$M \geqslant - Q(x) \qquad (4.1)$$

be fulfilled, where $Q(x) = a|x|^2 + b$ $(a,b \geqslant 0)$ or $Q(x)$ is constant in each of bodily closed layers of the sequence, which extend to infinity, and $Q(x) = \infty$ between them. On the other hand, in [25] the proof of the essential self-adjointness of the operator L of any even order, required $L \geqslant L_\varepsilon$, where L_ε contained derivatives of the same order as those in L, with an arbitrarily small coefficient $\varepsilon > 0$. But in addition to the equality $L^* = \overline{L}$, [25] presented a number of a priori estimates for elements of $D(L^*)$. It is intersting to know whether similar a priori estimates are valid under the condition (4.1). From a recent result of Brusentsev [31] it is follows, for example, that for the operator $L = (-\Delta)^m + q(x)$ the inequality

$$\|L\varphi\|^2_{L_2(R^n)} \geqslant \varepsilon \int_{R^n} |x|^{-2\frac{2m}{2m-1}} |\Delta^m \varphi|^2 \, dx, \quad (\varepsilon > 0)$$

guarantees both the essential self-adjointness and the validity of corresponding a priori estimates.

References

1. R.Arens: Operational calculus of linear relations, Pacif. J. Math. 11, No. 1, 1961, 9-23.
2. M.L. Gorbachuk, A.N. Kochubey, M.A. Rybak: Dissipative extensions of differential operators in a space of vector-function, DAN SSSR, 205, No.5 (1972), 1029-1032

3. M.L. Gorbachuk, A.N. Kochubey, M.A. Rybak: On some classes of extensions of differential operators in a spece of vector-function, in Collection "Primen. funkts. analiza k zadacham matem. fiziki", Kiev, 1973, 56-82.

4. A. Pleijel: A survey of spectral theory for pairs of ordinary differential operators, Lect. Notes Math., 448, Springer, (1975), 256-270.

5. F.S. Rofe-Beketov: Self-adjoint extensions of differential operators in a space of vector fuctions, DAN SSSR, 184, No.5 (1969), 1034-1037.

6. F.S. Rofe-Beketov: On self-adjoint extensions of differential operators in a space of vector functions: "Teor. Funktsiy, Funkts. Anal. i Prilozh.", vyp. 8, Kharkov (1969), 3-24.

7. V.I. Kogan and F.S. Rofe-Beketov: On square-integrable solutions of symmetric systems of differential equations of arbitrary order, Proc. Roy. Soc. Edinb., 74 A /1976/, 5-40.(There is a Russian preprint, FTINT AN Ukr.SSR, Kharkov, 1973).

8. V.I. Kogan, F.S. Rofe-Beketov: Precis of the article [7], R Zh Mat., 1976, 10Б612.

9. F.S. Rofe-Beketov, V.I. Kogan: On deficiency indices of systems with Nevanlinna type weight function. Proc. of the All-Union Math. Summer School, May 1975, Baku (to appear).

10. V.M. Bruk: On a number of linearly independent, square-integrable solutions of systems of differential equations, Funkts. anal., vyp. 5, Ulyanovsk, 1975, 25-33.

11. S.A. Orlov: Nested matrix circles, analytically dependent of parameter, and theorems on invariancy of ranks of limit matrix circle radii, IAN SSSR, ser. matem., 40, No. 3 (1976), 593-644.

12. E.S. Birger, G.A. Kalyabin: The Weyl circle theory for the case of nonself-adjoint systems of second-order differential equations, Dif. Ur., 12, No. 9 (1976), 1531-1540.

13. Bert Karlsson: Generalization of a theorem of Everitt. Proc. London Math. Soc. (2), 9 part I (Nov. 1974), 131-141.

14. Ch. Bennewitz: Generalization of Some Statements by Kodaira. Proc. Roy. Soc. Edinb. A 71, No. 2 (1973), 113-119.

15. I.M. Glazman: On some analog of extension theory of Hermitian operators and on nonsymmetric one-dimensional boundary-value problem on half-axis, DAN SSSR, 115, No. 2 (1957), 214-216.

16. F.V. Atkinson, W.N. Everitt and K.S. Ong: On the m-coefficient of Weyl for a differential equation with an indefinite weight function, Proc. London Math. Soc., (3), 29, (1974), 368-384.

17. O.F. Mikulina: Expansion in eigenfunctions of second-order differential equation with a single turning point. Dif. Ur., 7, No. 2 (1971), 244-260.

18. O.F. Mikulina: Expansion in eigenfunctions of polar-type second-order differential equation in partial derivatives for the case of whole plane. Dif. Ur., 10, No. 3 (1974), 550-552.

19. F.S. Rofe-Beketov: Spectral Analysis of Hill's operator and its perturbations, "Funktsional. analiz", vyp. 9, Ulyanovsk, 1977 (to appear).

20. F.S. Rofe-Beketov. Generalization of Prüfer's transformatio and discrete spectrum in the gaps of continuous spectrum. Proc. of the All-Union Mathem. Summer School, May 1975, Baku (to appear).

21. V.I. Khrabustovsky: On perturbation of the spectrum of arbi trary order self-adjoint differential operators with perio dic matrix coefficients. Matem. Fizika i Funktsional Anal., FTINT AN Ukr.SSR, vyp. 5 (1974), 123-140.

22. V.I. Khrabustovsky: On discrete spectrum of arbitrary order perturbed differential operators with periodic matrix coefficients. Mat. zametki, 21, No. 6 (1977), 829-838.

23. F.S. Rofe-Beketov, A.M. Kholkin: On relation between spectral and oscillation properties of the Sturm-Liouville matrix problem. Matem. sb., 102, No. 3 (3) (1977), 410-424.

24. A.G. Brusentsev: On spectrum of nonself-adjoint elliptic systems of arbitrary order. Dif. Ur., 12, No. 6 (1976), 1040-1051.

25. A.G. Brusentsev, F.S. Rofe-Beketov: Self-adjointness conditions for strongly elliptic systems of arbitrary order. Matem. Sbornik, 95, No. 1 (9) (1974), 108-129.

26. V.V. Grushin: To the proof of spectrum discreteness of some class of differential operators. Funkts. Anal. i Prilozh., 5, No. 1 (1971), 71-72 .

27. T. Kato: A remark to the preceding paper by Chernoff, J. Funct. Anal., 12, No. 4 (1973), 415-417.

28. P.R. Chernoff: Essential self-adjointness of powers of generators of hyporbolic equations. J. Funct. Anal., 12, No. 4 (1973) 401-414.

29. Yu.M. Berezanskiy: A notice on essential self-adjointness of operator powers. Ukr. Matem. zh., 26, No. 6 (1974), 790-793.

30. Yu. B. Orochko: Sufficient conditions for essential self-adjointness of polynomials of Schrödinger's operator. Matem. sb., 99, No. 2 (1976), 192-210.

31. A.G. Brusentsev: On J-self-adjointness of elliptic operators, satisfying no conditions of Titchmarsh-Sears. Matem. Fiz. i Funkts. Analiz, FTINT AN UkrSSR, vyp. 7 (to appear).

QUASI-SIMILARITY OF HILBERT SPACE OPERATORS

Béla Szőkefalvi-Nagy
Bolyai Institute, University Szeged
Szeged, Hungary

1. One of the basic notions in matrix theory is similarity: two (finite) square matrices, say A_1 and A_2, are called <u>similar</u> if there exists an invertible matrix X such that

(1.1) $$A_1X = XA_2, \text{ and hence, } A_2X^{-1} = X^{-1}A_1.$$

The most important classical result relative to this notion is that every square matrix over the complex field is similar to its canonical, or Jordan, form.

The notion of similarity immediately extends to infinite matrices, or more specifically, to (linear, bounded) operators on Hilbert space. However, if one tries to find, for some fairly general class of operators, a model more or less reminiscent to the classical canonical form, it turns out that similarity does not always work, indeed its role is taken over by some other relation, which in case of infinite dimensional spaces is weaker than (but in case of final dimensional spaces coincides with) similarity.

In order to define such a relation let us first introduce the notion of quasi-affinity. An operator X from a Hilbert space $\mathcal{H}_1$ into a Hilbert space $\mathcal{H}_2$ is a <u>quasi-affinity</u> (or <u>quasi-invertible</u>) if it is injective and "quasi-surjective", that is, if

(1.2) $$\ker X = \{0\}, \quad \overline{X\mathcal{H}_1} = \mathcal{H}_2 \quad \text{(bar denoting closure).}$$

Note that the second condition is equivalent to $\ker X^* = \{0\}$. Hence, if X is a quasi-affinity then so is X^*. Moreover, the product of two quasi-affinities $X: \mathcal{H}_1 \to \mathcal{H}_2$ and $Y: \mathcal{H}_2 \to \mathcal{H}_3$, is a quasi--affinity $YX: \mathcal{H}_1 \to \mathcal{H}_3$. In case of finite dimensional spaces, the second of the conditions (1.2) reduces to $X\mathcal{H}_1 = \mathcal{H}_2$; thus in this case every quasi-affinity is invertible (i.e., is an "affinity").

For two operators, say A_1 on $\mathcal{H}_1$ and A_2 on $\mathcal{H}_2$, we say that A_1 is a <u>quasi-affine transform</u> of A_2, and write $A_1 \succ A_2$ or $A_2 \prec A_1$, if there exists a quasi-affinity $X: \mathcal{H}_1 \to \mathcal{H}_2$ such that

$$A_2X = XA_1 .$$

If both $A_1 \succ A_2$ and $A_1 \prec A_2$ hold, i.e. if there exist quasi--affinities $X: \mathcal{H}_1 \to \mathcal{H}_2$ and $Y: \mathcal{H}_2 \to \mathcal{H}_1$ such that

$$A_2X = XA_1 \quad \text{and} \quad A_1Y = YA_2 ,$$

then A_1 and A_2 are called quasi-similar, and we write $A_1 \overset{q}{\sim} A_2$. (Note that in finite dimensional case the quasi-affinity X is an invertible operator so we can choose $Y = X^{-1}$; but in general the choice of X and Y is independent.)

Quasi-similarity is an equivalence relation, and moreover, $A_1 \overset{q}{\sim} A_2$ implies $A_1^* \overset{q}{\sim} A_2^*$.

Let us mention that for some particular types of operators, e. for normal operators A_1 and A_2, quasi-similarity implies unitary equivalence. It even suffices to suppose $A_1 \succ A_2$ only, i.e. that $A_1X = XA_2$ for some quasi-affinity. Indeed, by the Fuglede-Putnam theorem we have then $X^*A_1 = A_2X^*$, and hence, $A_2X^*X = X^*A_1X = X^*XA$ $A_2 |X| = |X| A_2$, where $|X| = (X^*X)^{1/2}$. From the fact that X is a quasi-affinity it readily follows that so is $|X|$, and that in the polar decomposition $X = V |X|$, V is a unitary operator. We deduce:

$$A_1V \cdot |X| = A_1X = XA_2 = V|X| A_2 = VA_2 \cdot |X| ,$$

and since $|X|$ has dense range we conclude that $A_1V = VA_2$, i.e. A_1 and A_2 are unitarily equivalent.

2. The concept of quasi-similarity first appeared in the paper [1], [2], in connection with contraction operators T on a Hilbert space $\mathcal{H}$, of class C_{11}, i.e. such that neither T^nh nor $T^{*n}h$ tend to 0 as $n \to \infty$, for any non-zero $h \in \mathcal{H}$. Indeed, it was proved that every such operator T is quasi-similar to some unitary operator. Later we noticed that the condition $\|T\| \leqq 1$ can be replaced here by "power-boundedness", i.e. $\|T^n\| \leqq M$ $(< \infty)$ for $n = 0, 1, \ldots$, and obtained ([H], Sec. II. 5):

Theorem. Every power-bounded operator T of class C_{11} is quasi-similar to a unitary operator U, indeed to a unique one, up to unitary equivalence.

Uniqueness follows from the fact that for unitary (indeed for all normal) operators quasi-similarity implies unitary equivalence.

In the proof of existence use is made of a "Banach limit" linear functional $L_{n\to\infty} x_n$ for bounded sequences $\{x_n\}$ of complex numbers. Indeed we start by defining, through

$$(2.1) \qquad \langle h,k \rangle = L_{n\to\infty}(T^nh, T^nk) \qquad (h, k \in \mathcal{H}),$$

hermitian bilinear form on $\mathcal{H}$; clearly satisfying

2.2) $$0 \le \langle h,h\rangle \le M^2\|h\|^2$$

nd such that (by the invariance property of $L_{n\to\infty}$)

2.3) $$\langle Th, Tk\rangle = \langle h, k\rangle \qquad (h, k \in \mathcal{H}).$$

hen there exists a self-adjoint operator X , $0 \leqq X \leqq M\,I$, such hat

2.4) $$\langle h, k\rangle = (Xh, Xk) \qquad (h, k \in \mathcal{H}),$$

nd, because (2.3) and (2.4),

2.5) $$(XTh, XTk) = (Xh, Xk).$$

Note that for any $h \in \mathcal{H}$ and for $n > m \geqq 0$ we have $\|T^n h\| = \|T^{n-m}T^m h\| \leqq M\|T^m h\|$, and hence,

$$\limsup_{n\to\infty} \|T^n h\| \leqq M \inf_{0\leqq m<\infty} \|T^m h\|.$$

ince $\|T^n h\|$ cannot tend to 0 for $h \neq 0$ we conclude that $\inf_m \|T^m h\| = \rho(h) > 0$ for $h \neq 0$. By (2.1) this implies, $\langle h, h\rangle \geqq \rho^2(h) > 0$ for $h \neq 0$, and hence, by (2.4), $Xh \neq 0$ for $h \neq 0$. Because X is self-adjoint this implies that X is a quasi-affinity on $\mathcal{H}$. From (2.5) it is clear that the map $Xh \mapsto XTh$ is isometric so it extends by continuity to an isometry U on $\overline{X\mathcal{H}} = \mathcal{H}$ such that

(2.6) $$UX = XT .$$

Actually, U is unitary on $\mathcal{H}$. To this effect, notice first that the condition $T^{*n}h \not\to 0$ for $h \neq 0$ implies $\ker T^* = \{0\}$, and therefore, $\overline{T\mathcal{H}} = \mathcal{H}$. Hence

$$U\mathcal{H} = U\overline{X\mathcal{H}} = \overline{UX\mathcal{H}} = \overline{XT\mathcal{H}} = \overline{X\overline{T\mathcal{H}}} = \overline{X\mathcal{H}} = \mathcal{H}.$$

Interchanging the roles of T and T^* we deduce in an analogous way that there exist on $\mathcal{H}$ a unitary V and a quasi-affinity Y such that $VY = YT^*$, $TY^* = Y^*V^*$. Combining this with (2.4) we get: $U \succ T \succ V^*$. As $U \succ V^*$ implies that U and V^* are unitarily equivalent we conclude that $T \overset{q}{\sim} U$.

Using a Banach limit functional $L_{t\to\infty}\, x(t)$ for bounded functions $x(t)$ of the real variable t on $[0,\infty)$ one can prove an analogous theorem for norm-bounded semigroups $\{T(t)\}_{0\leqq t<\infty}$ of operators, of class C_{11} , i.e. for which neither $T(t)h$ nor $T(t)^* h$ tend to 0 as $t\to\infty$, for any $h \neq 0$:

<u>Theorem.</u> <u>Every norm-bounded semi-group</u> $\{T(t)\}$, <u>of class</u> C_{11} ,

<u>of operators on $\mathcal{H}$ is quasi-similar to a semi-group $\{U(t)\}$ of unitary operators, indeed to a unique one, up to unitary equivalence.</u>

3. The following example is quite instructive, for it exhibits one of the main differences between similarity and quasi-similarity, namely that while similarity preserves the spectrum, quasi-similarity in general does not.

Let us start with an arbitrary selfadjoint contraction operator A on a Hilbert space $\mathcal{A}$, which is injective, i.e. $\ker A = \{0\}$. We construct the space

$$\mathcal{H} = \ell^2_{\mathcal{A}} \text{ of vectors } a = \langle \dots, a_{-1}, a_0, a_1, \dots \rangle, \; a_k \in \mathcal{A},$$

and we consider on $\mathcal{H}$ the operators:

(i) the bilateral shift U defined by $(Ua)_k = a_{k-1}$,

(ii) the operator $\tilde{A}$ defined by $(\tilde{A}a)_k = \begin{cases} Aa_0 & \text{for } k = 0, \\ a_k & \text{otherwise,} \end{cases}$

(iii) the operator $T = \tilde{A}U$.

Note that U is unitary, and $\tilde{A}$, T are contractions; $T^* = U^*\tilde{A}$.

It is immediate from the definitions that, for $n = 1,2,\dots,$

$$(U^{-n}T^n a)_k = \begin{cases} Aa_k & \text{for } -n \le k \le -1, \\ a_k & \text{otherwise,} \end{cases}$$

$$(U^n T^{*n} a)_k = \begin{cases} Aa_k & \text{for } 0 \le k \le n-1, \\ a_k & \text{otherwise.} \end{cases}$$

Hence,

$$\|T^n a\|^2 = \|U^{-n}T^n a\|^2 \to \sum_{-\infty}^{-1} \|Aa_k\|^2 + \sum_{0}^{\infty} \|a_k\|^2 ,$$

and

$$\|T^{*n} a\|^2 = \|U^n T^{*n} a\|^2 \to \sum_{-\infty}^{-1} \|a_k\|^2 + \sum_{0}^{\infty} \|Aa_k\|^2$$

as $n \to \infty$. Since A is injective, this implies $T \in C_{11}$.

Therefore, T is quasi-similar to a unitary operator. This turns to be actually equal our bilateral shift operator U. Indeed, it is immediate that the relations

$$TX = XU \quad \text{and} \quad UY = YT$$

hold in particular with the operators X and Y defined by

$$(Xa)_k = \begin{cases} Aa_k & \text{for } k \ge 0 \\ a_k & \text{for } k < 0 \end{cases}, \qquad (Ya)_k = \begin{cases} a_k & \text{for } k \ge 0 \\ Aa_k & \text{for } k < 0 \end{cases},$$

and both X and Y are quasi-affinities on $\mathcal{H}$ because so is A on $\mathcal{A}$.

Note that the "quasi-similar unitary model" U of T is independent of the particular choice of A, it only depends on the space $\mathcal{A}$, indeed U is the bilateral shift of multiplicity equal to $\dim \mathcal{A}$.

The spectrum $\sigma(U)$ is the unit circle $|z| = 1$. However, the spectrum $\sigma(T)$ can be different. Indeed we are going to show that if the point 0 belongs to $\sigma(A)$ (without being, as assumed, an eigenvalue of A), then (T) is the unit disc $|z| \leqq 1$.

Thus suppose $0 \in \sigma(A) \setminus \sigma_p(A)$.

For a fixed complex number z, $|z| < 1$, and for any $a \in \mathcal{A}$ we define $\tilde{a}(z) \in \mathcal{H}$ by

(3.1) $$\tilde{a}(z) = \langle \ldots, z^2a, za, a, \overset{\circ}{0}, 0, \ldots \rangle.$$

Then,

(3.2) $$\|\tilde{a}(z)\|^2_{\mathcal{H}} = (1 - |z|^2)^{-1}\|a\|^2_{\mathcal{A}}$$

and the operator T corresponding to A satisfies

$$T\tilde{a}(z) - z\cdot\tilde{a}(z) = \langle \ldots, z^2a, za, \overset{\circ}{Aa}, 0, 0, \ldots \rangle - \langle \ldots, z^2a, za, \overset{\circ}{0}, 0, 0, \ldots \rangle =$$

$$= \langle \ldots, 0, 0, \overset{\circ}{Aa}, 0, 0, \ldots \rangle ,$$

and hence,

$$\|T\tilde{a}(z) - z\cdot\tilde{a}(z)\|_{\mathcal{H}} = \|Aa\|_{\mathcal{A}} .$$

Since $0 \in \sigma(A)$, we can choose, for every $\varepsilon > 0$, an $a \in \mathcal{A}$ such that $\|Aa\| < \varepsilon \|a\|$. Then we have, also using (3.2),

$$\|T\tilde{a}(z) - z\cdot\tilde{a}(z)\|_{\mathcal{H}} < \varepsilon \|a\|_{\mathcal{A}} = \varepsilon(1 - |z|^2)^{\frac{1}{2}}\|\tilde{a}(z)\|_{\mathcal{H}} .$$

This implies that $z \in \sigma(T)$.

Thus we conclude that $\sigma(T) = \{z: |z| \leqq 1\}$.

The operator T considered in this example is essentially the same as the one considered in [H], See. VI.4.2, where it appeared as an application of the theory of characteristic functions.

Let us mention that there exist even quasi-similar operators A and B such that

$$\sigma(A) = \{0\} \text{ and } \sigma(B) = \{z: |z| \leq 1\},$$

cf. [5]. Of course, none of these is unitary.

4. Although the spectra of <u>quasi-similar</u> operators A, B need not coincide, there are still some loose connections between them.

Such a connection is that

$$(4.1) \qquad \sigma(A) \cap \sigma(B) \neq \emptyset , \qquad \text{cf. [3].}$$

A refinement of (4.1) was recently found by Fialkow [4], which we are going to reproduce, see (c) and (c′) below.

We rely on the Riesz-Dunford operator calculus for functions holomorphic on an (open, not necessarily connected) neighborhood of the spectrum. In what follows, A and B may mean operators on Banach spaces, say on $\mathcal{A}$ and $\mathcal{B}$ and X is an operator $\mathcal{B} \to \mathcal{A}$ such that

$$(4.2) \qquad AX = XB .$$

Then:

(a) <u>For any function</u> $f(z)$ <u>holomorphic on a neighborhood</u> $\mathcal{N}$ <u>of</u> $\sigma(A) \cup \sigma(B)$ <u>we have</u>

$$(4.3) \qquad f(A)X = Xf(B).$$

Indeed, if $\Gamma = \{\gamma_j\}$ is a (finite) set of contours in $\mathcal{N}$ encircling a (smaller) neighborhood of $\sigma(A) \cup \sigma(B)$ then (4.2) implies

$$X(zI_{\mathcal{B}} - B)^{-1} = (zI_{\mathcal{A}} - A)^{-1}X$$

for $z \in \Gamma$, and (4.3) follows when multiplying by $f(z)$ and integrating along Γ.

(b) <u>If</u> $\sigma(A) \cap \sigma(B) = \emptyset$ <u>then</u> $X = 0$.

By replacing A and B in (4.2) if necessary by $A - z_0 I$ and $B - z_0 I$ for some $z_0 \notin \sigma(A)$ we can suppose that $0 \notin \sigma(A)$. Then choose non-overlapping neighborhoods, $\mathcal{N}$ of $\sigma(A)$ and $\mathcal{M}$ of $\sigma(B)$, and consider the function $f(z)$ defined by

$$f(z) = z \text{ on } \mathcal{N}, \qquad f(z) = 0 \text{ on } \mathcal{M}.$$

By (4.3) $\qquad AX = X0 = 0, \quad X = A^{-1}AX = 0, \quad X = 0.$

(c) <u>If</u> X <u>is injective then every component</u> (i.e. non-empty, closed-and-open subset) σ <u>of</u> $\sigma(B)$ <u>satisfies</u>

$$\sigma(A) \cap \sigma \neq \emptyset .$$

Suppose the contrary, i.e. that for some σ we have $\sigma(A) \cap \sigma = \emptyset$. If $\sigma = \sigma(B)$ this would imply, by (b), that $X = 0$, contradicting injectivity of X. Thus we can assume that $\sigma \neq \sigma(B)$. Then take non-overlapping neighborhoods

$$\mathcal{N} \text{ of } \sigma(A) \cup \sigma , \quad \text{and} \quad \mathcal{M} \text{ of } \sigma(B) \setminus \sigma ,$$

and consider the function $f(z)$ defined by

$$f(z) = 0 \text{ on } \mathcal{N} \quad \text{and} \quad f(z) = 1 \text{ on } \mathcal{M}.$$

Then $f(A) = 0$, and $f(B)$ is a non-zero idempotent operator on $\mathcal{B}$.

By (a), $0 = f(A)\,X = X\,f(B)$, and since X is injective, $f(B) = 0$: contradiction. Thus (c) is proved.

(c′) _If_ X _is injective and_ $\mathcal{L}$ _is any non-zero invariant subspace for_ B, _then every component of_ $\sigma(B|\mathcal{L})$ _has a non-empty intersection with_ $\sigma(A)$.

One has only to apply (c) for $X|\mathcal{L}$ in place of X.

Some of these results also hold for the _essential_ spectra (i.e. the spectra of the corresponding elements in the Calkin algebra of operators modulo compact operators), _cf._ [4].

_5_. Thus, in contrast to similarity, which preserves spectra, quasi-similarity behaves rather poorly in this respect. It behaves somewhat better with respect to invariant subspaces. Indeed, one can show, _cf._ [H], Sec. II.5.1, that _if an operator_ A _has a non-trivial hyperinvariant subspace then so does every operator_ B _which is quasi-similar to_ A. (A subspace is hyperinvariant for an operator A if it is invariant for A as well as for all operators commuting with A.)

As unitary operators on a space of dimension higher than one do have hyperinvariant subspaces, this implies, by the theorem in Sec.2 that every power-bounded operator of class C_{11}, on a space of dimension > 1, has a non-trivial hyperinvariant subspace.

Quasi-similarity also preserves some lattice theoretic properties of hyperinvariant subspaces, cf. [H], Sec. II.5.

_6_. The most elaborate investigations done up to now involving the concept of quasi-similarity are those on the "Jordan model" of contractions of class C_0.

A contraction T on the Hilbert space $\mathcal{H}$ is called of class C_0 if (i) it is completely non-unitary, (ii) there exists a scalar valued inner function m such that $m(T) = 0$. Among these inner functions there is a minimal one, m_T, i.e. which is a divisor, in the algebra H^∞, of all the others.

For any given inner function m there exist operators $T \in C_0$ such that $m_T = m$, the simplest example being the operator $S(m)$ on the space $\mathcal{H}(m)$, defined by

$$(6.1) \qquad \mathcal{H}(m) = H^2 \ominus mH^2 , \quad S(m) = P_{\mathcal{H}(m)} S\,|\mathcal{H}(m) ,$$

where S denotes the simple unilateral shift on the Hardy space H^2 for the unit disc $|z| \leq 1$,

(6.2) $$S: u(z) \mapsto z\,u(z),$$

and $P_{\mathcal{H}(m)}$ is the orthogonal projection of the space H^2 onto its subspace $\mathcal{H}(m)$.

This operator is "multiplicity free", i.e. there exists a vecto $h \in \mathcal{H}(m)$ such that the the set $\{S(m)^k h\}$ $(k = 0,1,\dots)$ spans the whole space $\mathcal{H}(m)$, e.g. the vector

$$h = 1 - \overline{m(0)}m ,$$

and so is its adjoint $S(m)^*$. (Indeed, $S(m)^*$ is unitarily equivalen to $S(m^\sim)$, where

$$m^\sim(z) = \overline{m(\bar{z})} ,$$

the bar denoting here complex conjugate.)

The main result here is the following:

Theorem. Every operator $T \in C_0$ on a separable Hilbert space is quasi-similar to a unique operator of the form

(6.3) $$S(m_1) \oplus S(m_2) \oplus \dots \oplus S(m_k) \oplus \dots ,$$

where $m_1, m_2, \dots$ are inner functions each of which is a divisor of the preceding one.

The operator (6.3) is called the "Jordan model" of T for it has notable analogies to the Jordan (similarity) model (the "canonical form") of finite square matrices.

The theorem was proved, in [5], for the case that T has finite multiplicity μ_T ; this multiplicity if defined as the smallest cardinal of a set of vectors which, together with their transforms by T, T^2, etc. ..., span the whole space. In our case, the multiplicity μ_T turns out to be equal to the number of terms in the Jordan model (6.3) of T which are different from the trivial operator on the space $\{0\}$. Note that $\mathcal{H}(m) = \{0\}$ if and only if the inner function m is constant (of modulus 1).

The general case $\mu_T \leq \infty$ was proved by [6].

It is obvious that $m_1 = m_T$, but the proofs in [5] and [6] did not explicit the functions m_2, m_3, etc.

However, in the case that the contraction $T \in C_0$ has finite defect indices $d = \text{rank}\,(I-T^*T)$ and $d_* = \text{rank}\,(I - TT^*)$ (which are then automatically equal) so that the "characteristic function" $\Theta(z)$ of T is (cf. [H], chapter VI) a matrix of order d, over the Hardy-algebra H^∞ for the unit disc, then the functions $m_1, m_2, \dots$ can all be obtained from $\Theta(z)$ as the "invariant factors" of this matrix in the algebra H^∞. Cf. [7].

7. Every operator $T \in C_0$ has the properties that both T^n and T^{*n} strongly converge to 0 as $n \to \infty$; cf. [H], Chap. III.

The class of contraction operators T satisfying the only condition

$$T^{*n} \to 0 \quad \text{strongly, as } n \to \infty,$$

is denoted by $C_{.o}$.

For such operators there also exists a "Jordan model". More precisely, we have (cf. [8], [9], [10]):

Theorem. If $T \in C_{.o}$ and if, moreover, $I - TT^*$ has finite rank, then there exists a unique operator of the form

(7.1) $$J = S(m_1) \oplus S(m_2) \oplus \dots \oplus S(m_r) \oplus S \oplus S \oplus \dots,$$

where $m_i | m_{i-1}$ as above and S is the simple unilateral shift, so that J and T are related in the following way:

$$JX = XT \quad \text{and} \quad TY_i = Y_i J \qquad (i = 1,2)$$

for some quasi-affinity $X: \mathcal{H}_T \to \mathcal{H}_J$ and some injective operators $Y_i: \mathcal{H}_J \to \mathcal{H}_T$ $(i = 1,2)$ satisfying $Y_1 \mathcal{H}_J \vee Y_2 \mathcal{H}_J = \mathcal{H}_T$.

In general, J and T need not be quasi-similar, i.e. one cannot always find a single quasi-affinity

$$Y: \mathcal{H}_J \to \mathcal{H}_T \quad \text{such that} \quad TY = YJ .$$

Thus, in this case the relation between T and its Jordan model (7.1) is, in general, weaker than quasi-similarity. Indeed we are lead to a special case, in simplicity next to quasi-similarity, of the following general relation, which can be called "complete injection-similarity": two operators, say A on $\mathcal{A}$ and B on $\mathcal{B}$, are completely injection-similar if there exists a family $\{X_\alpha\}$ of injections $\mathcal{A} \to \mathcal{B}$ and a family $\{Y_\beta\}$ of injections $\mathcal{B} \to \mathcal{A}$ such that

$$BX_\alpha = X_\alpha A, \quad AY_\beta = Y_\beta B \quad \text{for all } \alpha, \beta ,$$

and

$$\bigvee_\alpha X_\alpha \mathcal{A} = \mathcal{B} , \qquad \bigvee_\beta Y_\beta \mathcal{B} = \mathcal{A} .$$

If we only require that at least one injection $X: \mathcal{A} \to \mathcal{B}$ and one injection $Y: \mathcal{B} \to \mathcal{A}$ exist with

$$BX = XA , \quad AY = YB ,$$

then A and B are called "injection similar". Note that both injection-similarity and complete injection-similarity are equivalence relations. For operator on finite dimensional spaces, both concepts reduce to similarity.

The implications of these weakened types of similarity are not yet fully understood; their investigation may prove to yield useful means for the analysis of the structure of operators.

REFERENCES

[H] B.Sz.-Nagy - C. Foiaş, Harmonic analysis of operators on Hilbert space, North Holland - Akadémiai Kiadó (Amsterdam - - Budapest, 1970).

[1] B.Sz.-Nagy - C. Foiaş, Propriétés des fonctions caractéristiques modèles triangulaires et une classification des contractions, C.R. Paris, 258 (1963), 3413-3415.

[2] B.Sz.-Nagy - C. Foiaş, Sur les contractions de l'espace de Hilbert. VII, Acta Sci. Math., 25 (1964), 38-71.

[3] T.B. Hoover, Quasisimilarity of operators, Illinois J. Math. 16 (1972), 678-684.

[4] L.A. Fialkow, A note on quasisimilarity of operators, Acta Sci. Math., 39 (1977).

[5] B.Sz.-Nagy - C. Foiaş, Modèle de Jordan pour une classe d'opérateurs de l'espace de Hilbert, Acta Sci. Math., 31 (1970), 91-115.

[6] H. Bercovici - C. Foiaş - B. Sz.-Nagy, Compléments à l'étude des opérateurs de classe C_0 . III, Acta Sci. Math., 37 (1975) 313-322.

[7] B. Moore III - E.A. Nordgren, On quasi-equivalence and quasi-similarity, Acta Sci. Math., 34 (1973), 311-316.

[8] B.Sz.-Nagy - C. Foiaş, Jordan model for contractions of class $C_{.0}$, Acta Sci. Math., 36 (1974), 305-322.

[9] B. Moore III - E.A. Nordgren, Remark on the Jordan model for contractions of class $C_{.0}$, Acta Sci. Math., 37 (1975), 307-312.

[10] B.Sz.-Nagy, Diagonalization of matrices over H^∞, Acta Sci. Math. 38 (1976), 223-238.

Scattering Theory for Partial Differential Operators

Joachim Weidmann

1. Introduction

Let T_1 and T_2 be selfadjoint operators in a complex Hilbert space H. The wave operators $W_\pm(T_2,T_1)$ are defined by

$$W_\pm(T_2,T_1) = \underset{t\to\pm\infty}{\text{s-lim}}\ e^{itT_2} e^{-itT_1} P_1$$

if these limits exist; here P_j is the orthogonal projection onto the absolutely continuous subspace $H_{ac}(T_j)$ of T_j,

$$H_{ac}(T_j) = \{ f \in H: \ \| E_j(.)f \|^2 \text{ is absolutely continuous with respect to the Lebesgue measure} \},$$

where E_j is the spectral resolution of T_j. $H_{ac}(T_j)$ is a closed subspace of H reducing T_j.

If the wave operators $W_\pm(T_2,T_1)$ exist, then

(a) $W_\pm(T_2,T_1)$ are partial isometries with initial set $H_{ac}(T_1)$ and final set $R(W_\pm(T_2,T_1)) \subset H_{ac}(T_2)$.

(b) The absolutely continuous part $T_{1,ac}$ of T_1, the restriction of T_1 to $H_{ac}(T_1)$, is unitary equivalent to the restriction of T_2 to $R(W_\pm(T_2,T_1))$; the unitary equivalence is given by $W_\pm(T_2,T_1)$.

(c) The equality $R(W_{\pm}(T_2,T_1)) = H_{ac}(T_2)$ holds if and only if the wave operators $W_{\pm}(T_1,T_2)$ exist also. In this case the wave operators are called complete. The wave operators $W_{\pm}(T_2,T_1)$ are then unitary operators from $H_{ac}(T_1)$ onto $H_{ac}(T_2)$.

(d) The scattering operator $S(T_2,T_1)$ is defined by

$$S(T_2,T_1) = W_+(T_2,T_1)^* \, W_-(T_2,T_1).$$

If the wave operators $W_{\pm}(T_2,T_1)$ are complete, then $S(T_2,T_1)$ is a unitary operator on $H_{ac}(T_1)$.

(e) If $W_{\pm}(T_2,T_1)$ are complete and $H_{ac}(T_1) = H$ (the last assumption is always true if T_1 is a non-trivial differential operator with constant coefficients in $H = L_2(\mathbb{R}^m)$), then $S(T_2,T_1)$ is a unitary operator in H. This property is expected in many important cases for physical reasons.

The aim of this paper is to prove the existence and completeness of wave operators $W_{\pm}(T_2,T_1)$, where T_1 is an elliptic differential operator with constant coefficients in $L_2(\mathbb{R}^m)$ (or slightly more general), and $V = T_2 - T_1$ is (something like) a differential operator of order less or equal to the order of T_1 with coefficients falling off at infinity like $|x|^{-m-\varepsilon}$. The results are applicable to Schrödinger and Dirac operators. We use only time dependent methods (see [8], section 11). The results are almost the same as earlier results by M.S. Birman [1] which were proved by means of both, time dependent and stationary methods.

It should be expected that the same results hold for a fall off condition like $|x|^{-1-\varepsilon}$. But so far we were not able to prove this with our method. However the existence of wave operators under such conditions has been proved in [6,7].

2. An interpolation result

We denote by B(H) the set of bounded and everywhere defined operators in H and by $B_\infty(H)$ the set of compact operators in H. For a compact operator A let $s_n(A)$ be the sequence of strictly positive eigenvalues of $|A| = (A^*A)^{1/2}$. For $1 \leq p < \infty$ we denote by $B_p(H)$ the set of compact operators in H satisfying $\sum_n s_n(A)^p < \infty$. The sets $B_p(H)$, $1 \leq p \leq \infty$, are ideals in B(H); they are Banach spaces with respect to the norms

$$\|A\|_p = (\sum_n s_n(A)^p)^{1/p} \quad \text{for} \quad 1 \leq p < \infty,$$

$$\|A\|_\infty = \|A\| = \sup \{s_n(A) : n \in \mathbb{N}\};$$

for proofs see [2], Section XI. 9, or [3], § III.7. $B_1(H)$ is called the _trace class_ of operators in H; $B_2(H)$ is the set of _Hilbert-Schmidt operators_ in H. We can now prove the following interpolation result.

Theorem 1. Let A be a bounded normal operator in H and V a closed operator in H with $R(A) \subset D(V)$ such that.

$$VA \in B(H) \quad \text{and} \quad VA^k \in B_1(H)$$

for some k = 2,3,.... Then for every $l \in \{2,3,\dots,k\}$ we have

$$VA^l \in B_{(k-1)/(l-1)}(H).$$

Proof. We first notice that without any restriction we may assume that A is a nonnegative selfadjoint operator. To prove this we use the fact that any bounded normal operator A can be written in the form $A = U|A| = |A|U$, where $|A| = (A^*A)^{1/2} \geq 0$ and U is a unitary operator. From

$$VA^1 = V|A|^1 U^1, \; V|A|^1 = VA^1 U^{-1}$$

it follows that VA^1 is in B(H) or in $B_p(H)$ if and only if $V|A|^1$ is in B(H) or in $B_p(H)$ respectirely.

We now assume that A is selfadjoint and $A \geq 0$. We consider the function

$$f: D \to B(H), \; z \mapsto V A^{1+(k-1)z},$$

where $D = \{ z \in \mathbb{C} : 0 \leq \text{Re } z \leq 1\}$. This function is strongly continuous on D and analytic on the interior $\overset{\circ}{D} = \{ z \in \mathbb{C} : 0 < \text{Re } z < 1\}$ of D. On the line Re z = 0 the function f is bounded with respect to the norm $\|\cdot\|$; on the line Re z = 1 it is bounded with respect to the trace norm $\|\cdot\|_1$ (both statements follow from the fact that $\|A^{it}\| = 1$ for real t).

We use now some complex interpolation theory. Let $X_o = B(H)$ and $X_1 = B_1(H)$. Then the interpolation spaces X_t, $0 \leq t \leq 1$, are given by

$$X_t = B_{1/t}(H) \quad (\text{with } \tfrac{1}{t} = \infty \text{ for } t = 0).$$

A proof of this result is indicated in [3], Chapter IX, Appendix 4, Prop. 8. Now the Calderón - Lions interpolation theorem ([4], Theorem IX. 2o) tells that

$$VA^s \in B_{(k-1)/(s-1)}(H) \quad \text{for } s \in (1,k].$$

This implies the above theorem. Q.E.D.

A more direct proof of this theorem can be given by means of the techniques of [3], Theorem 13.1.

3. The abstract scattering result

In [8], Satz 11.9 b), the author proved that the wave operators $W_{\pm}(T_2,T_1)$ exist if $VE_1(J) \in B_1(H)$ for every bounded interval J in $\mathbb{R}$. The completeness of the wave operators would follow if we would also have $VE_2(J) \in B_1(H)$ for every bounded interval J. But in general we are not able to prove that $VE_2(J) \in B_1(H)$ follows from $VE_1(J) \in B_1(H)$. Actually we need a slightly more restrictive condition. A similar result for semi-bounded operators was proved in [5].

__Theorem 2__. Let T_1 be a selfadjoint operator, V symmetric and T_1-bounded with relative bound < 1, $T_2 = T_1 + V$. If $V(z-T_1)^{-k} \in B_1(H)$ for some $z \in \rho(T_1)$ and some $k \in \mathbb{N}$, then $V(z-T_j)^{-k} \in B_1(H)$ for every $z \in \rho(T_j)$ (j = 1,2) and the wave operators $(W_{\pm}(T_2,T_1)$ exist and are complete.

For k = 1 and k = 2 this result has been proved in [8] by a much simpler method.

__Proof__. From $V(z-T_j)^{-k} \in B_1(H)$ for some $z \in \rho(T_j)$ it follows that for every bounded interval J

$$VE_j(J) = V(z-T_j)^{-k} (z-T_j)^k E_j(J) \in B_1(H),$$

since $(z-T_j)^k E_j(J)$ is bounded and $B_1(H)$ is an ideal in B(H).
Therefore it suffices to prove $V(z-T_2)^{-k} \in B_1(H)$.
Obviously $V(z-T_1)^{-1} \in B(H)$ and $V(z-T_1)^{-k} \in B_1(H)$ for every $z \in \rho(T_1)$. Since V has T_1-bound less than 1 we may choose $z \in \mathbb{C}\setminus\mathbb{R}$ such that

$$\| V(z-T_1)^{-1}\| = \varkappa < 1.$$

(Proof: By assumption $\|Vf\| \leq b\|Tf\| + a \|f\|$ with b < 1; hence for z = it we have $\|V(z-T_1)^{-1} f\| \leq b \| T_1(z-T_1)^{-1} f\| + a \|(z-T_1)^{-1}f \|$

$\leq (b + \frac{a}{t})\|f\|$.) This z is fixed for the rest of the proof.

Let now $R_j = (z-T_j)^{-1}$ for j = 1,2. From $\|VR_1\| = \varkappa < 1$ it follows that the expansion

$$R_2 = \sum_{n=o}^{\infty} R_1(VR_1)^n$$

holds (convergent in B(H)). This implies

$$VR_2^k = \sum_{l=1}^{\infty} \sum_{\substack{\alpha_o,\alpha_1,\ldots,\alpha_l \geq 0 \\ \alpha_o+\alpha_1+\ldots+\alpha_l = k-1}} VR_1^{\alpha_o+1} VR_1^{\alpha_1+1} V\ldots VR_1^{\alpha_l+1} .$$

If $\alpha_j = 0$, then

$$VR_1^{\alpha_j+1} = VR_1 \in B(H) \text{ and } \|VR_1^{\alpha_j+1}\| = \varkappa ;$$

if $\alpha_j > 0$, then

$$VR_1^{\alpha_j+1} \in B_{(k-1)/\alpha_j}(H) \text{ and } \|VR_1^{\alpha_j+1}\|_{(k-1)/\alpha_j} \leq C_1$$

with a constant C_1 independent of $\alpha_j \in \{1,2,\ldots,k-1\}$. Using the fact that for $A \in B_p(H)$, $B \in B_q(H)$ we have

$$AB \in B_r(H) \text{ and } \|AB\|_r \leq \|A\|_p \|B\|_q$$

with $\frac{1}{r} = \frac{1}{p} + \frac{1}{q}$ (see [3], § III. 7, Property 2), this implies

$$VR_1^{\alpha_o+1} VR_1^{\alpha_1+1} V\ldots VR_1^{\alpha_l+1} \in B_1(H),$$

and for $l \geq k$

$$\|VR_1^{\alpha_o+1} VR_1^{\alpha_1+1} V\ldots VR_1^{\alpha_l+1}\|_1 \leq C_2 \varkappa^{l-k}$$

with a constant C_2 independent of l and $\{\alpha_o,\ldots,\alpha_l\}$. For every l there are l^{k-1} terms of this form. Since the series $\Sigma \varkappa^{l-k} l^{k-1}$ converges, this implies the convergence of the above series for

VR_2^k in the trace norm $\|.\|_1$ (notice that $B_1(H)$ is a Banach space with respect to $\|.\|_1$). Therfore $VR_2^k = V(z-T_2)^{-k} \in B_1(H)$. Q.E.D.

We do not know whether the condition "V has T_1-bound less than 1" can be replaced by "V is T_1-bounded and $T_2 = T_1 + V$ is selfadjoint" In any case it is not difficult to prove the same result if V is only T_1-bounded and $T_1 + \lambda V$ is closed for every $\lambda \in [o,1]$ (compare [8], proof of Satz 9.2 for the technique of this proof).

4. Applications to partial differential operators

We prove first a general theorem about the existence and completeness of wave operators for operators in $L_2(\mathbb{R}^m)$. This will then be applied to partialdifferential operators.

Theorem 3. Let T_1 be a selfadjoint operator in $L_2(\mathbb{R}^m)$ such that $D(T^k) = W_{2,s}(\mathbb{R}^m)$ (the Sobolev space of order s) for some $k \in \mathbb{N}$ and $s > m$. Let further V be symmetric and T_1-bounded with T_1-bound less than 1, such that for some $p < s-m$

$$\| Vf \| \leq C(1+r)^{-m-\varepsilon} \| f \|_p, \qquad (*)$$

for every $f \in S(\mathbb{R}^m)$ with $f(x) = 0$ for $|x| < r$ (here $S(\mathbb{R}^m)$ is the Schwartz space and $\|.\|_p$ the Sobolev norm). Then with $T_2 = T_1 + V$ the wave operators $W_\pm(T_2,T_1)$ exist and are complete.

Remark. The above condition (*) holds if V is a differential operator of order less or equal p with coefficients falling off like $r^{-m-\varepsilon}$.

Proof of Theorem 3. It suffices to show that $VR_1^k \in B_1(H)$.

Let $\phi \in C_o^\infty(\mathbb{R}^m)$ be such that

$$\phi(x) \geq 0, \ \int\phi(x)dx = 1, \ \phi(x) = 0 \text{ for } |x| \geq \tfrac{1}{2}.$$

For every $\gamma = \{\gamma_1,\ldots,\gamma_m\} \in \mathbb{N}^m$ let

$$Q_\gamma = \{x \in \mathbb{R}^m : \gamma_j \leq x_j \leq \gamma_j + 1\},$$

$$Q_\gamma = \{x \in \mathbb{R}^m : \gamma_j - 1 \leq x_j \leq \gamma_j + 2\},$$

$$\phi_\gamma = \int \phi(x-y)\chi_{Q_\gamma}(y)dy = \phi * \chi_{Q_\gamma},$$

$$\hat{\phi}_\gamma = \phi * \chi_{Q'_\gamma} .$$

The family $\{\phi_\gamma : \gamma \in \mathbb{R}^m\}$ represents a partition of unity. The maps

$$\phi_\gamma R_1^k : L_2(\mathbb{R}^m) \xrightarrow{R_1^k} W_{2,s}(\mathbb{R}^m) \xrightarrow{\phi_\gamma} W_{2,p}(\mathbb{R}^m)$$

are in $B_1(L_2(\mathbb{R}^m), W_{2,p}(\mathbb{R}^m))$, since $R_1^k : L_2(\mathbb{R}^m) \longrightarrow$ $\longrightarrow W_{2,s}(\mathbb{R}^m)$ is bounded and $\phi_\gamma : W_{2,s}(\mathbb{R}^m) \longrightarrow$ $\longrightarrow W_{2,p}(\mathbb{R}^m)$ is in $B_1(W_{2,s}(\mathbb{R}^m), W_{2,p}(\mathbb{R}^m))$. We have $\|\phi_\gamma R_1^k\|_1 \leq C_1$ with some constant C_1. By assumption the maps

$$V\phi_\gamma : W_{2,p}(\mathbb{R}^m) \longrightarrow L_2(\mathbb{R}^m)$$

are bounded with $\|V\phi_\gamma\| \leq C_2(1+\gamma)^{-m-\varepsilon}$ for some constant C_2. This implies that the series

$$VR_1^k = \sum_\gamma (V\phi_\gamma)(\phi_\gamma R_1^k)$$

converges in $B_1(L_2(\mathbb{R}^m)$ and therfore $VR_1^k \in B_1(L_2(\mathbb{R}^m)$. Q.E.D.

Example 1. Let $T_1 = -\Delta$ in $L_2(\mathbb{R}^m)$ with $D(T_1) = W_{2,2}(\mathbb{R}^m)$. Let V be a symmetric differential operator of the form

$$Vf = \sum_{|\alpha| \leq 2} c_\alpha D^\alpha f,$$

where

$$\sum_{|\alpha|=2} \sup \{|c_\alpha(x)| : x \in \mathbb{R}^m\} < 1 \qquad (**)$$

and

$$|c_\alpha(x)| \leq C_\alpha(1+|x|)^{-m-\varepsilon} .$$

(The fall off conditions for $\alpha < 2$ may be weakened to fall off conditions in the sense of Stummel mean values.) Then V is T_1-bounded with relative bound less than 1; $T_2 = T_1 + V$ is selfadjoint. The wave operators $W_{\pm}(T_2,T_1)$ exist and are complete. To prove this we have to choose the number k in the above theorem such that $2k > m + 2$.

Example 2. Let T_1 be the unperturbed Dirac operator in $L_2(\mathbb{R}^3)^4$ with $D(T_1) = W_{2,1}(\mathbb{R}^3)^4$. Then the corresponding result holds if V is a symmetric differential operator of order less or equal 1 in $L_2(\mathbb{R}^3)^4$. In this case we have to choose k such that $k > m + 1$. In the above examples $H_{ac}(T_1)$ was equal to $L_2(\mathbb{R}^m)$ or $L_2(\mathbb{R}^3)^4$ respectively. Therefore the results imply that $W_{\pm}(T_2,T_1)$ are isometries and $S(T_2,T_1)$ is unitary. Theorem 3 can also be applied to differential operators T_1 with non-constant coefficients; in this case $H_{ac}(T_1)$ is not necessarily the whole space. We give only a simple example to illustrate such applications.

Example 3. Let $T_1 = -\Delta + S$ in $L_2(\mathbb{R}^m)$, where

$$Sf = \sum_{|\alpha|\leq 2} d_\alpha D^\alpha f$$

with smooth coefficients d_α such that (**) holds and the derivatives of d_α up to the order $m + 2$ are bounded. Then for k with $m + 2 < 2k < m + 4$ we have $D(T^k) = W_{2,2k}(\mathbb{R}^m)$. This makes it possible to apply Theorem 3 if V is as in Example 2 with

$$\sum_{|\alpha|=2} \sup \{ |c_\alpha(x)| + |d_\alpha(x)| : x \in \mathbb{R}^m \} < 1.$$

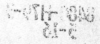

References

[1] Birman, M.S.: Scattering theory for differential operators with constant coefficients. Functional Anal. Appl. 3 (1969) 597 - 61o

[2] Dunford, N.; Schwartz, J.T.: Linear operators Part II: Spectral theory, Selfadjoint operators in Hilbert space. New York / London 1963

[3] Gohberg, I.C.; Krein, M.G.: Introduction to the theory of non-selfadjoint operators in Hilbert space. Providence, R.I. 1969

[4] Reed, M.; Simon, B.: Methods of modern mathematical physics II : Fourier analysis, selfadjointness. New York - London 1975

[5] Reed, M.; Simon, B.: The scattering of classical waves from inhomogeneous media. To appear in Math. Zeitschrift

[6] Veselić, K.; Weidmann, J.: Existenz der Wellenoperatoren für eine allgemeine Klasse von Operatoren. Math. Zeitschrift 134 (1973) 255 - 274

[7] Veselić, K.; Weidmann, J.: Asymptotic estimates of wave functions and the existence of wave operators. J. Funstional Analysis 17 (1974) 61 - 77

[8] Weidmann, J.: Lineare Operatoren in Hilberträumen. Stuttgart 1976